The Alzheimer's Workbook

Holistic Health and Problem Solving
for Everyday Care

Elizabeth Cochran, R.N., Ph.D., Th.D.

Publisher:
Communications Arts
P.O. Box 10423
Albuquerque, NM 87184

Website: Alzheimersworkbook.com

Cover design by Alena Cochran
Text design by Melanie Wegner
Edited by Virginia Stephens

Library of Congress-in-Publication Data

Cochran, Elizabeth
Alzheimer's Workbook, Holistic Health and Problem Solving for Everyday Care

ISBN 978-0-578-02598-8
1. Alzheimer's Disease. 2. Dementia. 3. Caregiving. 4. Holistic Health.

Printed in the United States of America

A Blessing

To those of you who have
Chosen to take on the special
Charge of caring for someone
With Alzheimer's disease.
Your caregiving is a spiritual gift
To an individual and to the world.
By your actions, you are the antidote
To the helplessness and hopelessness
That Alzheimer's brings into the world.
From those who have gone before
Our hearts and prayers are with you always

ACKNOWLEDGEMENTS

I wish to acknowledge the special people who have contributed to the accomplishment of this book. First I must truly dedicate this book to my mother-in-law, Mary Cochran (pictured in the cover art) who was a loving grandmother whose full life was drawn short by Alzheimer's disease. She was a treasure and provided an education to my entire family. My wonderful husband Jim who has been my copilot and project manager in getting the workbook finished, without his help all of this would not have been possible. My delightfully artistic daughter Alena who designed the cover in memory of her grandmother and to my other children Jonas and Melissa who have been great support through the many years it took to create and publish the Workbook. I will love them all for eternity.

Melanie Wegner for her wondrous artistic skills in designing the format for this book, Virgina Stephens who sifted through the sands of my writing to edit and polish it. My sister Anna Harwood who sorted and organized a large part of my research. My friend Marie Kerns who is always there to listen.

The New Mexico Alzheimer's and Related Disease Association, and the support group leaders who so kindly allowed me to attend their meetings and enlist participants for my dissertation study using the first draft of the Workbook. I would also like to acknowledge the help of the dedicated people of Share Your Care in Albuquerque who helped with enlisting participants for the dissertation study.

My heartfelt appreciation and thanks to the chair of my dissertation committee, Dr. Ann Nunley for her enduring help and support as well as the members of my committee, Dr. Norman Shealy, Dr. Robert Nunley, and Dr. Robert Matusiak. I have truly appreciated the opportunity to learn from each of them and participate in this wonderful experience that is Holos University Graduate Seminary.

Last but not least, thanks to my great (times 5) grandfather, the great scientist and statesman Benjamin Franklin who lives on in family legend and serves as inspiration for me in my desire to innovate and be of service to others.

TABLE OF CONTENTS

SECTION I

SECTION II

TABLE OF CONTENTS
continued

INTRODUCTION

This workbook was born out of my own personal frustration and continuing struggle to learn to be a caregiver and advocate for my mother-in-law who had Alzheimer's disease. Her descent into Alzheimer's disease and the sudden need to move her into our home was stressful for our whole family. My husband had to accept that she did not recognize him anymore. My children and I had to deal with the loss of her kind, caring personality.

As a "health professional", I might have seemed to have it easier; but I was just as angry and frustrated as any other person dealing with Alzheimer's disease/dementia (AD). I tried to become more informed by reading books and articles. Most of them I pushed away half read. Some books were just too complex and others were just too wordy, with the practical suggestions lost in the text.

After I learned about the basic theories of the disease, I realized that there are many definitions of the stages of AD, each with its accompanying behaviors. I discovered that these stages were difficult to apply, because AD is very different from a progressive disease like cancer where there are well-defined indicators of a worsening condition. With AD, problem behaviors come and go. Just when you are sure the person has lost the ability to do something, the knowledge seems to pop back into their heads (at least for a short time).

I discovered that the majority of the problems that occurred with my mother-in-law were with her behaviors and ideas such as, putting clothes on backwards, insisting dead parents are coming to dinner, being angry over a lost lipstick, or wandering at night. These are not clearly defined medical problems; therefore I didn't often have definite reasons to call her physician. Thus I was left with my own judgment and not much advice. I sought out support and advice in the community as well as formal support groups. But it was hard finding practical suggestions as fast as I needed them. So, I learned to guess a lot, trust my instincts, and depend on trial and error. This process was stressful but not negative because I learned to problem solve. I now believe that many of the best and most practical suggestions come from caregivers. Taking all this into consideration, I wondered if there wasn't a better way to find help in working with AD related behavior problems.

I enlisted the help of friends and experts in the field of aging and AD to explore the idea of writing a resource book. This workbook was created out of my personal need for answers and my fascination with the creativity of caregivers. I broadened my scope to include suggestions for helping the caregiver cope, maintain sanity, and learn to look more clearly at problems and potential solutions as well as learning to BE with a person with AD. Beyond all the day-to-day issues, persons with AD never stop being human, which includes needing to give and receive love/care as well as to be useful even is some small way. Life never seems to stop needing meaning.

I decided that designating three simple stages of the disease seemed most useful:

1. Diagnosis and the Early Times when you find out your loved one has AD and when there are only minor losses.

2. Middle Stage (or Middle Madness) which includes all the years that the person retains the ability to perform some of the activities of daily living, but you think YOU are going to lose your mind.

3. Late/End Stage when all functions have to be provided for the person and death is not far off.

The hardest stage (physically and emotionally) is the Middle (Madness) stage. The span of this stage covers many accumulating losses and can represent up to 20 years of care. Many of us would like more firm milestones, but each individual seems to lose skills at his/her own pace. We can predict the next skill to change or be lost but not exactly when it will happen. Knowing which skill may be the next to change or be lost is very helpful. The behaviors list breaks down behaviors and skills into various levels to help you in making these predictions, and the workbook gives you ideas to prepare for coming declines.

In creating a format for the workbook, I chose to focus on wellness. It may seem odd at first for a chronically ill person to be considered well. Most people think of Alzheimer's disease/dementia (AD) as just a chronic illness, but many people with AD have few medical problems. They are more disabled than ill and many of them are actually quite well, physically. It is true that eventually they will become nonfunctional but this can take many years. Because of this, focusing on wellness makes sense. The wellness concept is a way of looking at the whole person, accentuating the positive, and working with a person's strengths. Encouraging wellness may seem like a very hard task but it can be easier than you think. Some of the problems resolve themselves and occasionally chronic complaints can gradually disappear. My mother- in-law's stomach problem faded away when she could no longer remember what she was constantly worrying about. Other problems require thought and work. For instance, by satisfying a person's social needs through community programs such as adult day care, you can improve the quality of life, increase wellness, and preserve some of the self-esteem of the person with AD. In general, a relatively well person with AD is easier to handle than one who has physical/medical problems along with AD.

There are five wellness areas: **1) mental, 2) physical, 3) emotional, 4) social** and **5) spiritual.** All are very important when looking at the whole person as in this holistic health approach. The behaviors and problem-solving ideas have been organized under these five wellness areas. The lists of behaviors are presented in a simple format for easy skimming. One purpose of this workbook is ease of use in a crisis, when you need quick ideas and not a lot of explanations. You can start right in with Chapter 1, **How to Use This Book** to learn the problem-solving system. The later informational chapters were written to prepare you for problem solving and caring for your loved one and yourself. You can read these chapters whenever you feel like it. The chapter on "**retrogenesis**" or reverse development **(Chapter 8)** is also key to creating healthful strategies. It is my sincere hope that this workbook will ease your life. After you have read and used the workbook, please let me know if it has helped you or if you have some successful caregiving strategies or suggestions for future printings (contact information is at the back of the book).

CHAPTER 1
How to use this book

Hopefully by now you have read the introduction and your interest has been piqued. This how-to section has been placed before the longer discussion chapters to allow you to get right to practical problem solving. The rest can be saved until you have time for a thoughtful read.

The main body of the book is divided into the five-wellness/whole person categories: **physical, emotional, mental, social** and **spiritual.** The behaviors presenting the largest potential caregiving problems are listed under these five wellness categories. The behaviors have been broken down into specific areas, going from positive behaviors (easiest to handle) to major losses (hardest to handle). The Behaviors Log is also designed with writing space for keeping track of your person's behaviors, the solutions you have tried, and ones that are currently working. When you make these written notes, the log can be shown to your person's physician and passed to other family members or surrogate caregivers. The Behaviors Log can help you estimate in general terms how disabled your person with Alzheimer's disease/dementia (AD) is becoming as they begin to lose more of their skills. The list has been coded to show areas of full function (hatched areas), problem losses (gray areas) and severe losses (black areas). In general, the farther down the list a person is , the more serious the loss and the more disabled s/he is. The Behaviors discussions and coping strategies (Chapter 3) is your key to problem solving for coping ideas.

STEPS for using this book for problem solving:

1. Decide on the problem behavior/skill that you want to work on.

2. Go to the Behaviors log: Chapter 2, page 9.

3. Find the wellness area the behavior/skill falls under: (Physical, Emotional, Mental, Social or Spiritual)

4. Go down the list until you find the behavior/skill that you want to work on.

5. Find the page number listed in the far right column next to the problem behavior and go to that page.

6. On that page you will find a short description of the behavior, followed by some suggested ideas for you to try. The suggestions are arranged from the easiest to the hardest for you to use and for the person with AD to tolerate.

7. Pick the idea that seems to work best for you and the person you are caring for. Remember there are no right or wrong answers, just ones that work or don't work for you and the person you are caring for. Mark that page and make notes in the margins.

8. Give the idea you picked a try. Give it some time to work. If one of the ideas sparks an idea of your own, try it. Remember some of the best ideas come from caregivers.

9. Go back the idea you picked. Circle and date the suggestions you tried and make some notes. This creates a very valuable record.

10. Use the workbook for all your problems and areas of concern. Reading ahead is valuable to give you more ideas to try and to prepare for future issues.

STEPS for using the Behaviors log to keep a record of your person's abilities:

1. Go to the Behaviors log. Start at the beginning and mark under each category the behavior that your person currently exhibits and the date. The hatched, gray and black markings in the left column indicate whether the behavior is in the early, middle or late stage for each behavior (see example below). This gives you a quick way to see where your person with Alzheimer's disease/dementia (AD) is in retaining or losing that skill or behavioral ability and where s/he might be in overall stages of AD. This also creates a valuable record for your family, your helping network, and you person's physician.

STAGES	YES √DATE	BEHAVIORS LOG	PAGE
EARLY		Can read labels and signs	
MIDDLE		Can read some words - Poor attention span	
LATE		Cannot read	

2. Some behavior areas will still be positive or healthy. In the behaviors suggestions (Chapter 3) some ideas to help preserve strengths and perhaps enhance enjoyment are given.

3. Circle the number of any suggestions you are trying. Use extra space to make notes about the behavior or problem. You can refer to the helpful suggestions, as you need them. You don't have to work it all out at once.

4. Feel free to write in this book. Mark or tab your important pages. It is a WORKBOOK designed to be a tool to help you cope.

CHAPTER 2
Behaviors log

STAGES	YES √DATE	BEHAVIORS LOG	PAGE
		🧠 **MENTAL**	
		📖 READING	23
		Enjoys reading books	23
		Enjoys reading magazines	24
		Can read labels and signs	24
		Cannot read labels and signs	24
		Cannot read	25
		✏️ WRITING	25
		Can write	26
		Can use computer	26
		Can write short message	26
		Can write name only	27
		Can not write	27
		➗ MATH	27
		Can do complex math	27
		Can balance checkbook or accounts	27
		Can do simple math	28
		Cannot use math	28
		💲 MONEY	28
		Can use money correctly	29
		Understands purchasing and can make change	29
		Cannot independently handle money but wants to make purchases	29
		Cannot manage money but opposes your assistance	30
		Cannot handle money and no longer cares	30
		🧠 IMAGINING AND HALLUCINATIONS	30
		Has no problem with imagining or hallucinations	31
		Imagines and converses with deceased relatives	31
		Imagines that certain events have taken place or that things are missing	31
		Imagines doubles(two bedrooms or kitchens)	32
		Imagines other homes and wants to get there	32

STAGES	YES √DATE	BEHAVIORS LOG	PAGE
		Becomes very upset over these imaginings	32
		Actively engages hallucinations/imaginings	32
		📝 INSTRUCTIONS	33
		Can follow a verbal list of several tasks	33
		Can follow visual instructions for tasks	34
		Can do two or three tasks one after another	34
		Can only do one task at a time	34
		Must be supervised for the whole task	34
		Cannot understand instructions	34
		🕐 TIME	35
		Can tell time	35
		Can't tell time but enjoys watches and clocks	35
		Time is irrelevant	36
		🕰 POSSESSIONS	36
		Can handle and organize personal possessions	36
		Hoards possessions	36
		Loses possessions	37
		Loses possessions and blames others	38
		Doesn't care about possessions	38
		🗃 REMINISCENCE	38
		Loves/likes to review the past	38
		Indifferent about the past	39
		Destroys objects from the past	39
		🕴 **PHYSICAL**	40
		👓 EYE SIGHT	40
		Can see without difficulty	40
		Must use glasses	40
		Complains about vision in spite of glasses	40
		Has difficulty with vision tests	41
		Has cataracts	41
		Has cataracts that need surgery	41
		Legally blind (not late stage if natural)	41
		👂 HEARING	42
		Has no hearing difficulty	42
		Complains of hearing difficulties	42

STAGES	YES √DATE	BEHAVIORS LOG	PAGE
		Hearing impaired	43
		Deaf (but not late stage if natural)	43
		👃 SMELLING	43
		Can smell without difficulty	43
		Reports things that do not smell as good as they used to smell	43
		Uses too much perfume or cologne	44
		Diminished appetite	44
		Seems unresponsive to smell	44
		👄 TALKING	44
		Speaks with little difficulty	44
		Uses words incorrectly or cannot find the right word	45
		Difficulty making sentences	45
		Makes little or no sense when speaking	46
		Nonverbal and uses gestures to express self	46
		Cannot speak and does not respond to conversation	46
		Randomly shouts or cries out	47
		📞 TELEPHONE	47
		Can use telephone	47
		Can take messages	48
		Cannot take messages	48
		Cannot dial or hang up phone	48
		Cannot use the phone at all	48
		🍴 EATING	49
		Has no difficulty eating	49
		Has difficulty using utensils	50
		Hides food in napkin or hoards food	50
		Tries to feed pets from plate	51
		Needs assistance with eating food	51
		Has difficulty chewing or swallowing	52
		Unable to feed self	52
		Needs to be fed in bed	53
		Cannot swallow	53
		🍽 APPETITE	53
		Has good appetite	54

STAGES	YES √DATE	BEHAVIORS LOG	PAGE
		Has large appetite	54
		Has small appetite	54
		Has no appetite	55
		🥛 NUTRITION	55
		Vitamins	57
		Eats well balanced meals	58
		Eats unbalanced meals	58
		🍖 PROTEINS	58
		Likes to eat protein foods	59
		Protein foods enjoyed-*list on page*	59
		Has difficulty eating protein foods	59
		🍞 CARBOHYDRATES	60
		Likes to eat complex carbohydrates-*list on page*	60
		Complex carbohydrates enjoyed	61
		Likes mostly simple carbohydrates	61
		Simple carbohydrates enjoyed-*list on page*	62
		Has a passion for sweet foods	62
		Eats very few carbohydrates	62
		🍲 FATS	62
		Likes to eat fats	63
		Fats allowed-*list on page*	63
		Eats very few fatty foods	63
		🥦 VEGETABLES	63
		Likes to eat vegetables	63
		Vegetables enjoyed-*list on page*	63
		Vegetables hated-*list on page*	64
		Refuses to eat most vegetables	64
		🍇 FRUITS	64
		Likes to eat fruits	64
		Fruits enjoyed-*list on page*	64
		Fruits hated-*list on page*	64
		Refuses to eat most fruits	65
		🥤 FLUIDS	65
		Likes to drink water	65
		Likes to drink coffee	65

STAGES	YES √DATE	BEHAVIORS LOG	PAGE
		Likes to drink fruit juice	65
		Doesn't like to drink fruit juice	65
		Doesn't like to drink water	66
		🌏 ALCOHOL	66
		Does not drink alcoholic beverages	66
		Likes to drink alcoholic beverages	66
		Has a chronic drinking problem	67
		🌍 SMOKING	67
		Does not smoke but is ex-smoker	67
		Enjoys smoking and can manage it	68
		Cannot hold cigarette or cigar properly	68
		Does not want to quit smoking	68
		Family members smoke	68
		Has lung disease	69
		👐 ARMS AND HANDS	69
		Able to grasp and carry things	69
		Has weak grasp but can hold or carry things	69
		Cannot hold or carry things, has little use of hands and arms	70
		🚶 WALKING	70
		No difficulty walking	70
		Difficulty walking	70
		Shuffles	71
		Stumbles	71
		Tendency to fall	72
		Difficulty walking up or down stairs	72
		Needs to use cane	73
		Needs to use walker	73
		Needs help getting in or out of bed	73
		Needs help getting in or out of chair	73
		Cannot walk spontaneously or independently	74
		Needs to use wheelchair	74
		Requires transfer from wheelchair to bed or chair	74
		Is bed bound	74
		🧭 WANDERING	75
		Tendency to wander	75
		Wanders at night	76

STAGES	YES √DATE	BEHAVIORS LOG	PAGE
		🐵 TRANSPORTATION & DRIVING	77
		Can drive	77
		Able to use a bus or taxi	77
		Cannot safely drive but will not give up driving	78
		Must be driven places	78
		🐊 EXERCISE	78
		Likes to exercise on a routine basis	79
		Seldom exercises	79
		Likes to play sports	79
		Recreational activity of choice	80
		Has no interest in exercise or sports	80
		Movements are severely restricted	80
		🏠 MANEUVERING IN THE HOUSE	81
		Able to negotiate the house and recognize rooms	81
		Unable to recognize or find own room	81
		Unable to recognize other rooms in house	81
		Wanders into other family members' rooms	81
		Cannot recognize outside of house	82
		Imagines there are other rooms or another floor in the house	82
		➕ SAFETY	82
		Safety is not yet an issue	82
		Beginning to use household appliances unsafely	83
		The house needs to be safety /dementia proofed	83
		🔍 COOKING	84
		Can cook without help	84
		Can cook with help	84
		Can turn on and off, stove and oven	85
		Can turn on stove or oven, but forgets to turn off	85
		Can pour boiling water	86
		Cannot pour boiling water	86
		Can read recipe and measure ingredients	86
		Cannot read recipe and measure ingredients	86
		Can use microwave	87
		Cannot use microwave	87

STAGES	YES √DATE	BEHAVIORS LOG	PAGE
		Able to set table	87
		Cannot manage table setting at all	87
		Can use small electrical appliances	88
		Cannot use small electrical appliances	88
		Can slice and chop with knife	88
		Must be supervised when slicing or chopping	88
		Cannot perform any cooking tasks	88
		🦪 WASHING DISHES	89
		Can wash dishes without help	89
		Can wash and dry dishes with supervision	89
		Puts away dishes in wrong places	90
		Rearranges kitchen drawers or cabinets	90
		Can tell the difference between leftover food and trash	90
		Cannot tell the difference between leftover foods and trash	90
		Tends to toss good food or utensils in the trash	90
		Cannot perform any washing tasks	90
		🧴 CLEANING	91
		Cleans without difficulty	91
		Can clean with supervision	91
		Cleans excessively	92
		Can use vacuum cleaner	92
		Can use vacuum cleaner with supervision	92
		Cannot use vacuum cleaner	92
		Can sweep and use dust pan	93
		Cannot sweep and use dustpan	93
		Can wipe counters and table	93
		Can wipe counters and table with supervision	93
		Can empty trash	93
		Cannot perform any cleaning tasks	93
		🧺 WASHING CLOTHES	94
		Can use washer and dryer	94
		Can use washer and dryer with supervision	94
		Cannot use washer and dryer	94
		Can do hand washing	95
		Can do hand washing with supervision	95

STAGES	YES √DATE	BEHAVIORS LOG	PAGE
		Can fold laundry	95
		Can put away laundry	95
		Can fold laundry with supervision	95
		Can put away laundry with supervision	96
		Cannot fold laundry properly	96
		Can sort socks	96
		Can sort socks with supervision	96
		Cannot sort socks	96
		Cannot perform any laundry tasks	96
		⚒ WORKING AROUND THE HOUSE/SHOP	97
		Can work in shop/garage or make household repairs without help	97
		Can work in shop/garage or make household repairs but needs supervision.	97
		Cannot perform any shop/garage or household repair tasks	98
		🛠 WORKING AROUND THE YARD	99
		Can perform outdoor yard tasks without help	99
		Can perform outdoor yard tasks but needs supervision.	100
		Cannot perform any outdoor yard tasks	101
		🌱 GARDENING	101
		Can perform gardening tasks without help	101
		Can perform gardening tasks but needs supervision	102
		Cannot perform any gardening tasks	103
		👕 DRESSING SELF	103
		Dresses self appropriately	104
		Tends to put clothes on backwards or mismatches outfits	104
		Tends to reverse use of clothes (shirts for pants)	104
		Occasionally cannot find clothes in closet or drawers	105
		Cannot recognize own clothing	105
		Tends to wear too many clothes	105
		Tends to wear too few clothes	105
		Likes to go naked sometimes	106
		Puts shoes on the wrong feet	106

STAGES	YES √DATE	BEHAVIORS LOG	PAGE
		Cannot find socks and shoes	106
		Cannot tie shoes	106
		Cannot put shoes on	107
		Cannot dress self at all	107
		⚪ ACCESSORIES	107
		Can find purse, wallet, or other valuable items	107
		Often cannot find purse or wallet	108
		Often cannot find glasses	108
		Often cannot find hat and gloves	109
		Worries that accessories have been stolen	109
		Often hides personal items	109
		Can put on coat and zip or button it	109
		Cannot button or zip coat	110
		Cannot manage coat	110
		Cannot manage any accessories	110
		⚪ PERSONAL HYGIENE – BATHING	110
		Bathes self without assistance	110
		Can take sponge bath by self	111
		Can take tub bath by self	111
		Can take shower by self	111
		Often forgets and must be told to bathe	112
		Afraid to use tub or shower	112
		Afraid of water	112
		Needs assistance bathing	113
		Must be given a full bath	113
		⚪ TOOTH BRUSHING	113
		Brushes teeth or dentures by self	114
		Needs reminder to brush teeth or dentures	114
		Occasionally uses wrong cleaner for teeth or dentures	114
		Handles dentist visit well	114
		Doesn't handle dental visit well	114
		Needs help cleaning dentures	115
		Cannot recognize toothpaste or toothbrush or needs help with tooth brushing	115

STAGES	YES √DATE	BEHAVIORS LOG	PAGE
		Cannot brush teeth at all	116
		🦷 GROOMING	116
		Can brush or comb hair by self	116
		Can shave self	117
		Can apply own make-up	117
		Needs help with care for hair	117
		Needs help to shave	117
		Needs help to apply make-up	118
		Uses too much cologne or perfume	118
		Must be shaved	118
		Requires total assistance with grooming	118
		🚽 TOILET-Urine	119
		Can hold urine and toilet self	119
		Has an occasional toileting accident	119
		Needs help with toileting	120
		Incontinent of urine only at night	121
		Incontinent of urine	122
		🚽 TOILET- Bowels	123
		Can manage own toileting with bowel movements	123
		Has diarrhea	124
		Is prone to constipation	124
		Has an occasional accident	125
		Incontinent of bowel movements	125
		🧦 SKIN & FOOT CARE	126
		Can care for own skin and skin is in good shape	126
		Skin dry	127
		Foot and toenail care are needed	127
		Has cuts, wounds or lesions	127
		Cannot perform own skin and nail care	127
		🌙 SLEEP	128
		Has no sleep difficulties	128
		Has difficulty falling asleep	128
		Paces at night	128
		Wanders at night (sundowning)	129

STAGES	YES √DATE	BEHAVIORS LOG	PAGE
		⊙ SEX	130
		Has interest in sex and can perform intercourse	130
		Has a sexual partner and desires intercourse	130
		Does not have a sexual partner and desires intercourse	131
		Cannot perform intercourse but wants to	131
		Makes inappropriate sexual advances	131
		Makes inappropriate social advances	132
		Removes clothing and tends to expose self	132
		Masturbates publicly	132
		Has no interest in sex	132
		⊙ **EMOTIONAL**	133
		⊙ FEELINGS	133
		Able to express appropriate emotions	133
		Has to be pushed to talk about feelings	134
		Expresses inappropriate emotions	134
		Has catastrophic reactions	134
		⊙ MOODS	135
		Happy and cheerful most of the time	135
		Occasionally sad or depressed	136
		Often sad, depressed, or cries a lot	136
		Expresses loneliness and depression	136
		Talks about self-destruction/suicide	137
		Is mostly non-responsive	137
		⊙ MENTAL ILLNESS	138
		Has clinical mental illness	138
		Has clinical depression	138
		Has talked about suicide	139
		⊙ ANGER	139
		Expresses appropriate anger	139
		Expresses inappropriate anger	139
		Has violent reactions	140
		⊙ WORRY, RESTLESSNESS AND AGITATION	141
		Occasionally fretful and worried	141
		Excessively fretful, worried, and restless	142

STAGES	YES √ DATE	BEHAVIORS LOG	PAGE
		Paces a lot	142
		Difficulty with self quieting	143
		◯ OBSESSIONS	143
		Has no obsessions	143
		Hoards possessions and takes others items	143
		Jewelry and fancy accessories	144
		Money	144
		Clothing	145
		Food	145
		Other items	145
		Takes other people's possessions	146
		◯ PARANOIA	146
		General	146
		🌐 SOCIAL	147
		◯ RELATIONSHIP WITH FAMILY	147
		Able to recognize family members	147
		Able to intermittently recognize family members	147
		Responds well to a particular family member	148
		Responds poorly to certain family members	149
		Often is rude and unthankful to family	149
		Routinely unable to recognize family members	149
		Does not remember visits from family	150
		Unresponsive to family members	150
		◯ RELATIONSHIP WITH FRIENDS	151
		Able to recognize friends	151
		Able to intermittently recognize friends	151
		Responds best to a particular friend	152
		Has friends that live out of area	152
		Unhappy that friends never visit	153
		Responds poorly to certain friends	153
		Often is rude and unthankful to friends	153
		Routinely unable to recognize friends	153
		Unresponsive to friends	154
		◯ OTHER SOCIAL RELATIONSHIPS	154
		Able to be social in groups	154

STAGES	YES √DATE	BEHAVIORS LOG	PAGE
		Has difficulty being social in groups	155
		Will not participate in group activities	155
		Makes inappropriate social advances	155
		Unresponsive in social situations	156
		🐑 SOCIAL BEHAVIORS	156
		Often is pleasant and polite to acquaintances	156
		Often is snappy or rude to acquaintances	156
		🔴 VISITING	156
		Can go for visits to relative or friends houses	157
		Often tries to take other people's possessions	157
		👥 RESPITE	157
		Can go for overnights with friends or relatives	157
		Can handle extended stays with friends or relatives	158
		Becomes agitated or restless or periodically becomes a behavior problem	158
		🌍 RECREATIONAL ACTIVITIES	159
		Enjoys recreational activities	159
		Enjoys work activities as recreation	160
		Favorite recreational activities-*list on page*	160
		Does not enjoy recreational activities	161
		Refuses to participate in recreational activities	161
		🌀 VACATIONING	162
		Can handle vacation trips	162
		May be able to travel alone	163
		Unable to fly alone but still able to fly	163
		Unable to go on vacation with you and/or your family	165
		Unresponsive to vacation settings	165
		🙏 **SPIRITUAL**	165
		⭐ BELIEFS	165
		Believes in God, Supreme Being, or Universal Divine State	165
		Setting up spiritual sessions	166
		Accessing and working in a spiritual state	167
		Does not believes in God, Supreme Being or Universal Divine State	168
		Follows a certain religion	169

STAGES	YES √DATE	BEHAVIORS LOG	PAGE
		Has a good relationship with a clergyman/ woman/spiritual teacher	169
		Does not remember how to behave during religious services	169
		Has no religious or spiritual interest	170
		♥ SPIRITUAL COMMUNICATION	170
		Person with AD and caregiver share the same religious denomination	170
		Both are religious but do not share the same denomination	171
		Person with AD is religious and caregiver is spiritual/nondenominational	171
		Person with AD is spiritual/nondenominational and caregiver is religious	172
		🕊 DEATH	172
		Expresses feelings about death	172
		Very worried or scared about death	173
		Upset over deaths of friends or relatives	173
		Will not or cannot talk about death	173
		You have the opportunity to be present at the death of your person with AD.	173

CHAPTER 3
Behaviors discussions and coping strategies
MENTAL
READING

General

Reading is a lifelong activity for many people. Try and keep reading behavior active for as long as you can. The person's ability to read is also a good indicator of how s/he is doing in general. Reading is an activity that will enhance a person's brain activity. Even the act of pretending to read can be familiar and comforting.

Enjoys reading books

If your person enjoys reading or has in the past, it is an activity that you should encourage. Reading is relaxing, comforting, familiar and feels normal. Your person does not have to understand or retain what is read to enjoy the act of reading. Keeping your person actively reading can provide you with needed time to accomplish your own tasks. Or you can enjoy quiet time reading together.

• If there is a certain kind of books that your person likes, have the books available.

• Library books can be a great resource in early stages, but watch out for any tendency to forget and rip out pages. You will not want to replace costly library books.

• Shop in second-hand stores. They often have quite a collection of mysteries, science fiction, and westerns. If these books are destroyed, you are not out much money.

• Be delicate with conversations that require recounting of stories or material that have recently been read. Reading retention is easily lost. Reading is important by itself.

• Encourage reading time over TV time if that suits your lifestyle.

• Make sure you set up some time for reading each day. The routine may be helpful to your person and it will give you much needed respite.

• Don't overlook children's books. They are colorful and easy to read. If you are not sure if your person is interested in or ready for this material, buy or borrow some and leave them out where your person can see them and read them. If they are cast aside, save these books because they could be helpful in the future.

• Encourage reading stories out loud, if it is comfortable to do so.

• Eventually your person will just be going through the motions of reading the book. S/he may even read it upside down. Don't correct this. Just leave your person alone in the enjoyment of whatever s/he is seeing.

• Try books on tape. Your person may be able to listen and understand a story after the ability to read is lost.

Notes:

Enjoys reading magazines

Magazines are much more of a quick read, no plots to keep track of. Many people find this kind of reading relaxing. If this has been something your person enjoys, keep it going. The reading time will provide you with respite.

- Keep supplied with the magazines that your person likes. If you are not sure, try a few different kinds.

- Borrow previously read magazines from friends and family when they are finished with them.

- Buy the popular grocery store variety; they are fairly cheap and have short stories and pictures to keep the mind occupied.

- Visit a bookstore that has lots of magazines and allows people to read them. Be careful that your person does not forget and begin damaging the magazines.

- Subscribe to magazines your person likes.

- Subscribe to magazines that you both like or share the ones you like. At some point the content of the magazine won't matter that much.

- Your person will begin to forget what s/he reads and old magazines will seem new again. This can cut down on having to get new ones.

- Look for big-print magazines created for those with vision difficulties.

- Look for more picture-oriented magazines. These can be less stressful than ones with a great deal of print

- Don't overlook children's magazines. They are colorful and easy to read. If you are not sure if your person is interested in or ready for this material, buy or borrow some and leave them out where your person can see them and read them. If they are cast aside, save these magazines because they could be helpful in the future.

- Eventually a person with Alzheimer's disease/dementia (AD) will just be going through the motions of reading the magazine. S/he may even read it upside down. Don't correct this. Just leave your person alone in the enjoyment of whatever s/he is seeing.

Can read labels and signs

So far the discussion has been about recreational reading, but situational reading is very important to your person's ADL (activities of daily living). The reading of signs and labels is very important for safety in establishing where you are or what you are going to eat. You will want to know when your person loses this skill, because at that time s/he will need an increased level of supervision.

- Encourage the occasional reading of general signs out loud when you both go out.

- Encourage the occasional reading of road signs as you drive.

- Encourage the occasional reading of food labels

Cannot read labels and signs

At this point you have some safety issues that you need to be concerned with.

- If your person cannot read road signs, s/he should not be driving because s/he can easily get lost or make a dangerous traffic error.

- Walking must be supervised. Do not let your person go out of the house alone. S/he will most likely get lost. You do not want the stress of

hunting for hours or having to call in a report to the police.

• Obtain a Medic Alert or some other safety bracelet. This will help people who find your person to return him/her safely to you.

• Participate in the Alzheimer's Association's Safe Return program. (Call your local branch.)

• Childproof your home. This may sound strange but your person is regressing to childhood abilities and may not be able to recognize the difference between food and poison. The first taste may stop your person, but at some point (just like small children) s/he can get a fair amount down before deciding it doesn't taste good. This constitutes putting away all poisonous materials in secure or locked cabinets. You can consider buying childproof devices for the kitchen cabinets. Most of these devices are sold in the baby departments in grocery or department stores.

• If you are not sure if your childproofing efforts are making sense, talk to someone who has a small child and get their advice on the most useful ideas and devices.

• You may want to keep your local poison control number handy, just in case your person eats something and you are not sure if it is harmful.

• Do not forget common things like mothballs, Clorox, or kitchen cleansers.

• Supervise any activity that requires the reading of labels or instructions.

Cannot read

Reading is such a conditioned behavior. Your person may not know s/he can no longer read, or it may be distressing when the words don't make sense. Distress can be minimized by simply encouraging the act of reading. It just matters that the experience be enjoyable.

• Don't force or embarrass your person with insistence on reading.

• Continue to have books and magazines available in case your person wants to try to read. This can still be a routine and comforting activity.

• Children often take advantage of situations where adults seem less competent. Make certain children don't tease your person.

• Get your person books on tape. Load and unload the tape player if your person cannot.

• Read to your person or have your children or grandchildren read to him/her. This can be a source of pleasure for everyone.

• Remember and refer to the safety issues around not being able to read signs and labels.

WRITING

General

We take writing for granted. We've done it in one way or another all our lives, but there is more to writing behavior than meets the eye. First you have to have the motor skills to manipulate the pen or pencil, or type on the keyboard. You need the cognitive skills for the part that requires organizing thoughts, and then there are language and grammar skills for placing those thoughts in proper order and syntax. Depending on the level of proficiency the person once had for writing, losing these skills can be

Notes:

Notes:

more or less of a problem. A person who only wrote shopping lists and quick notes will miss writing less than one who did it for a living.

Can write

As with all basic skills, it is a good idea to preserve them as long as you can. This prevents rapid onset of loss of function and your person will be easier to care for.

- If your person wrote for a living, encourage that behavior or even the simulation of that behavior for as long as possible.

- If s/he liked long hand, keep pad and pencils/pens handy.

- Don't worry if the writing begins to make less and less sense as long as your person is enjoying him/herself.

- Encourage your person to write personal stories and experiences. This will be valuable for posterity.

- Encourage your person to write letters or emails to friends and extended family. Help him/her monitor the correspondence and to write back as soon as an answer arrives.

Can use a computer

This is certainly a modern skill, but more and more people over 60 are acquiring computing skills. If your person can still use a computer, encourage him/her to do so. There will come a time when this will no longer be possible and computing may become a source of frustration. Until then:

- If your person can type, a computer keyboard may be a good way to get down his/her thoughts. Encourage your person to write on the computer whenever s/he feels like it.

- If s/he becomes confused about certain parts of the task, fill in the blanks. Post cue cards for process instructions.

- Turn the computer on and off for your person.

- Make the system as simple as possible to use. Enlist the help of your computer literate children or people in your support network.

- Encourage your person to write out his/her life story. This can be important for posterity.

- If your person knows how to use email, encourage him/her to write to friends and relatives and to read their responses.

- If computing becomes uncomfortable or frustrating, shut off the computer and encourage your person to write in any way that is comfortable.

- If your person keeps trying to go back to the computer and is getting upset, remove the whole set-up to another room where you can use it or hide it until your person no longer cares about it. The disease will eventually make him/her forget.

Can write short message

This could be an area where writing skills and organizing skills can be reinforced. In early stages, list and note making can be used as a reminder system.

- Encourage your person make lists of things s/he needs.

- If you need to leave a note when you both go out, have your person do it.

Notes:

• Have your person write reminder notes on post-its and place them in appropriate places. This can be used for appointments or parts of tasks that are easily forgotten.

Can only write name

This is a middle stage issue. Your person has regressed to a childlike ability and now must be supervised in this area. By this time, you have probably taken care of financial issues. If not, you will need the help of an attorney.

• Monitor those things your person signs, since it's a leftover behavior s/he may want to sign anything s/he can or is asked to do.

• Make sure, at this stage, if there are things that require your person's signature you get it done while s/he can still sign.

Cannot write

This is a late stage issue and not too much of a practical problem as long as all the financial matters have been taken care of.

• Encourage your person to draw or make marks if s/he likes working with pencil and paper.

• Move your person in the direction of art and drawing if s/he is still longing to work with paper and pencils.

• Let him/her make a mark if s/he wants to sign something.

MATH

General

Math is a skill that seems certainly linked to the type of intelligence your person may have. The scope of reaction to math ranges from genius to hating it. If your person has some proclivity to math it may be a skill you will want to preserve. If your person disliked math, your only issues will be those of finance. If you have the needed power of attorneys, then math will not be too much of a problem. Losses in the area of math are often seen early on the mini mental assessment for Alzheimer's disease/ dementia (AD). Your person can seem quite functional in other areas, but be unable to balance his/her checkbook and be headed for financial trouble.

Can do complex math

If your person could do complex math or was involved in a profession that needed high-level skills, your person may still enjoy looking at facts and figures. Since it is one of the skills last acquired it may be one that erodes first. This is less likely to happen if she has used the skill consistently over the years.

• Encourage your person to work with or look over facts and figures that s/he dealt with in his/her job or profession.

• Do this as long as it brings your person joy and a sense of involvement.

• When your person becomes stressed over matters, then phase this activity out. Without the stimulation, your person will forget the importance of the activity as well as how to do it.

Can balance checkbook or accounts

Even in the early stages of AD, your person's math skills are very likely impaired or will be. If you can, double check his/her checkbook figures and

Notes:

the bills s/he is paying. As well as items they may be ordering from the TV or catalogues. You do not want any nasty financial surprises. Ability with math is a good indicator of his/her progress with AD.

• This skill can erode rather quickly and it is dangerous.

• It is ok to encourage your person to do as much as possible for him/herself. You want to keep him/her mentally active. Just be watchful.

• You may need to find ways to offer your help gently as managing finances may be a self esteem issue with your person.

• Read the section on Money and make sure everything is in order before his/her accounting skills are gone.

Can do simple math

If your person enjoys working with simple figures, then let them do so. You can make it a game and something to talk about. If it is annoying, drop the whole area and find some other skill to encourage.

• At the grocery store, for instance, encourage your person to tell you the prices of things.

• If you rely on your person to do any calculations at all, double-check what s/he tells you.

• If your person takes phone messages, that involve times or amounts, double-check.

• Let your person help you with cooking, especially measurements.

• Let your person help you with counting items, for instance have him/her get you half a dozen oranges when you're grocery shopping.

Cannot use math

Except for finances this skill can fade and be accommodated. Just understand and accept the person with dementia abilities and try not to frustrate them.

• Once your person cannot do math at all, s/he should be handling only small amounts of money.

• You need to have all financial matters under your control.

• Be careful not to burden them with making them count, as in "peel me six potatoes."

• Be patient with them when dealing with anything that involves counting or adding.

• Make instructions as simple as possible.

MONEY

General

Money can be a difficult issue. Unless you have been handling your person's financial affairs for a period of time, introducing the transfer of financial control can be very touchy. Money is closely tied to one's sense of self and self worth. A person's sense of freedom is often rooted in his/her ability to spend money the way he or she wants. Depending on the extent of the patient's estate and income, there may be the potential for significant conflict with family members or other interested parties, i.e. business partners, creditors, etc. Once diagnosed, it is best to start the process of closely monitoring your person's financial affairs. You should also seek legal advice to obtain the necessary power of attorney to be used when it becomes clear that s/he can no longer safely manage money. Also consider your person's vulnerability to exploitation, which can occur because your person does not understand the consequences of their agreements. This includes mortgages, gimmicky sale items, long term purchasing agreements, and even religious donations. You do not want to deprive a person of their independence prematurely, but this is an area of great danger if not monitored very carefully.

Can use money correctly

Be sensitive to your person's insecurity and fears when talking money, but begin the discussion of finances.

- Try and inventory the finances.
- Observe how your person handles money and credit cards.
- Pay close attention to how they are handling their own bills. If your person recognizes that math has become difficult or if s/he has never been comfortable with it, this may be an opportunity to step in and offer support.
- Discuss the future while your person has "lucidity" and can understand what is needed for safe handling or his/her finances.
- Seek legal advice on obtaining necessary "power of attorney" for later

Understands purchasing and can make change

If you start to notice some slipping in ability to make purchases and especially in tracking money, then you need to slowly begin to offer support. Often the losses are deeper than what is observable. Your person may be more confused than you think. Alzheimer's disease/dementia (AD) may also be affecting his/her emotional responses and your person may become more paranoid or resistant to change. Diplomacy is important. Try not to get into an argument, but slowly work around his/her logic, giving supportive and protective reasons to make the needed changes.

- Your person should have some cash and access to some money, but this is the stage to start limiting it.
- At the first sign of confusion, start limiting the use of credit cards.
- Discuss a budget and use of limited amounts of cash
- Discuss taking over the checkbook and bill paying.
- If you take over any of the finances be very careful and keep good records. You do not want to be questioned by family members and have no support for purchases you have made with another's money.

Cannot independently handle money but wants to make purchases

When your person starts to have problems with money, it is an indication that some alternative financial arrangements should be made. If you haven't obtained a power of attorney yet, try and get it done now. You have to start thinking seriously about where your person may start "leaking " money. Persons with AD fall easy prey to telephone solicitation. Prepare yourself for the loss of some money. It may be through forgotten purchases or simply misplaced ones.

- Many phone companies offer solicitation-screening messages for a nominal charge. This may be the time to have that installed.

Notes:

• Your person should have some cash and access to some money, but limit it. Giving cash to your person in small bills will make it feel like more money.

• Whatever amount you give your person, be sure it is an amount you can afford to lose.

• Bankcards may work, but s/he probably will not be able to remember the password. You also don't want your person handing the card to strangers and asking for help. And you don't want to write the pin number on the card.

• Consider pre-written checks to certain frequently used vendors or stores. Some stores may also accept small-prepaid accounts or allow you to keep a tab. Keep all of this low key, so there is little chance of exploitation.

• Allow your person to make purchases under your supervision where you provide whatever level of support needed. This way you can preserve your person's dignity.

• At this point, credit cards should be withdrawn from use.

Cannot manage money but opposes your assistance

This is a tricky phase. You may have missed some of the more lucid moments when your person could be easily convinced to accept help.

• Continue to observe for moments of lucidity, reason, or a moment when your person recognizes money is getting hard for them to manage. Begin your discussions then. Be solicitous, offering comfort and protection. Do not emphasize inadequacy or humiliation; this will only make your person angry and more resistant.

• Slowly begin taking over the finances and avoid head-on arguments.

• Try to get the power of attorney forms signed in a moment of lucidity, otherwise, you may find yourself having to go for officially declaring your person incompetent and becoming his/her guardian. Guardianship is a complicated and expensive process and more of a last resort.

• If all your papers are in place and your person is just being oppositional, try to give them a number of small bills so that your person can feel that it is the money s/he wanted or needed.

• You may have to fib a bit and give them less money than s/he asks for. It is very likely s/he will not be able to count it. It should be an amount that you both can afford to lose.

Cannot handle money and no longer cares

By now you have completely taken over your person's finances; give them small bills if it makes them feel better.

IMAGINING AND HALLUCINATIONS

General

Reality is relational. It is based on our interpretation of what we perceive. Our conceptual mental framework accommodates the ideas that people are either living or dead. Time is past, present or future. Objects have a specific use. Geography is based on time and distance. For persons with Alzheimer's disease/dementia (AD) these distinctions gradually deteriorate. As a result, they begin to perceive the world differently than they did

Notes:

before. It is a changing landscape for them and they strive to give order to these altered perceptions. Their interpretations can often lead to novel, entertaining and yes, dangerous situations. Understanding what's coming is helpful for the family and caregivers. The perceptual difficulty persons with AD experience are often interpreted as hallucinations. Since their mental state is often complicated by other physical impairments such as hearing loss, poor eyesight (further complicated by losing glasses or hearing aids) these perceptual difficulties can be mere misinterpretations unmediated by their waning judgment. It is possible that this mental state can also be representations of internal psychological and spiritual activity. In general it is best to <u>work with</u> your person's perceptions. You will have to learn to let loose of your attachment to "reality" and play along. If your person thinks it is Tuesday, then let it be Tuesday even if it is Friday. You will have much more peace in your life if you do not argue about these matters.

Has no problem with imagining or hallucinations

• If your person has been diagnosed with Alzheimer's disease/dementia (AD), this is the time to get a complete mental and physical examination. This will give you a snapshot in time of your person's health and give you an indication of any underlying conditions, either mental or physical that may add to their confusion as the disease progresses.

• Treat your person normally, but listen carefully and speak clearly.

Imagines and converses with deceased relatives

This is a frequent occurrence and has to do with time displacement. It can be unnerving for your mother to be having a conversation with her deceased mother. What often accompanies this condition is the belief that parents are still alive and that people such as children or a spouse is actually someone else. A grown son may be viewed as his father, an uncle, or older relative. In your person's frame of reference, **retrogenesis** (reverse development, **CHAPTER 8**) progresses s/he sees him/herself as actually being younger. While frustrating and often the source of much sadness at not being recognized by someone you've loved and known all your life, it is just the natural course of the disease and you should refrain from trying to argue them out of their ideas.

• Don't confront or argue with your person about this.

• Be patient and comforting. Listen. Take your person into family memories and use this to learn some things about the family's past.

• Talk about what is happening and see if s/he knows why this is happening. Follow the train of thought; see where it leads. Continue to be of comfort.

• If your person becomes overly excited or experiences deep sadness, slowly try and distract him/her with another activity.

Imagines that certain events have taken place or things are missing

Many persons with AD suffer from delusions that things have been taken, or are missing, or will experience a feeling of an impending event triggered by an old memory.

• Remain calm and let him/her talk it through.

• If it is an event that s/he is worried about missing or having missed,

Notes:

talk about what is happening and see if s/he knows why this is happening. Follow the train of thought to see where it leads. Assure your person that things can be rectified. Continue to be of comfort.

• If s/he is talking about missing items, offer assurances that you will look for them.

• Empathize with their loss. "I know what it's like when I can't find things, but I'll look for it and let you know when I find it."

• Keep an extra set of keys, an extra wallet or handbag for such occasions. Producing these, if your person with dementia has actually hidden the object, will reassure them. Then try and find the real thing.

Imagines doubles (two bedrooms or kitchens)
My mother-in-law moved in with us from the East. She was used to two-story houses and often wanted to go upstairs. We would simply walk her around through the living room and back to the kitchen and she would be satisfied. It is amazing what a bit of clever imagination will do.

• Create a short ritual like the one listed above to let your person believe s/he has arrived in the desired place.

• Tell your person you will take him/her later.

• Give some simple direction and let your person try to find the place in the house. When the person returns, telling him/her that they have arrived.

• Provide distraction. (see next section)

Imagines other homes/places and wants to get there
This can be a dangerous situation. The person believes s/he is going to his/her "other" home or back to their original home. This leads to random wandering and can result in them getting lost or hurt.

• Supervise or gently confine the person's ability to wander.

• Try a short walk around the block where you come back to the "other " house. This will serve to distract your person, and the physical exercise will help calm them.

• Distract to another activity or a snack.

Becomes very upset over these imaginings
One of the curses of this disease is also one of the blessings. Persons with AD have a hard time remaining focused when there are other things vying for their attention. When your person is agitated, first try and calm them down. Logical arguments may not work but empathetic emotional persuasion can work wonders.

• Don't engage their upset. Your anger/anxiety is only going to escalate your person's agitation. Exercise and distraction are two very good ways of taking your person's mind off of things.

• Ask for his/her help with a small chore like looking through a magazine for pictures to cut out for a project that you want to do.

• Offer a snack or a cookie, this will change their blood chemistry rapidly and may have a calming effect.

Actively engages hallucinations/imaginings
It is common in the middle and later stages of AD for people to develop

Notes:

visual, auditory or tactile hallucinations. These involve seeing, hearing or feeling things that don't seem to be there. The changes that AD induces in the brain can cause a myriad of perceptual aberrations, and in the moment of experiencing a hallucination, it will certainly seem real. Visual hallucinations are often caused by misinterpretation of arrangements of objects or patterns, much like seeing an image in a cloud or on the ceiling. These are often enhanced by sensory difficulties. Persons with AD also tend to develop delusions, beliefs that people or circumstances are other than what they are, despite evidence to the contrary.

• If you gently show your person evidence that his/her imaginings are just not justified and they remain unconvinced, remain calm and don't argue. Try to avoid becoming the object of your person's agitation.

• If hallucinations are neither destructive nor dangerous, just remain calm and talk about the experience. Act as if the circumstance is real; it is for your person.

• Remember, for the moment, <u>this is reality for your person</u>. Engage the hallucination a bit: "Sure the horse is on the lawn, but we'll call animal control and get it off…would you like some cookies?" While persons with AD are often impervious to logic, they are very sensitive to emotion; so stay calm, even and loving; and most of all try and remain respectful within the context of the experience the person is having.

• If your person has imaginings about him/herself such as that s/he is a nurse and this is a hospital, and if your person is no threat to anyone, play along. Avoid ridicule sarcasm and irony.

• Gently distract them to some safer activity.

INSTRUCTIONS
General
There are a number of factors that will influence your person's ability to follow instructions. If there is a history of love and trust in the relationship, things will go easier. Another important factor is the degree to which your person has progressed in the disease. It is important that you as the caregiver exercise patience. As the disease progresses, so does your person's disorientation. His/her sense of time, place and even self-recognition begins to deteriorate. As orientation fails, your person will begin to falter in his/her ability to listen to and perform a series of tasks. And eventually, s/he will not understand how to follow instructions at all. We often take this ability for granted and when your person begins to fail, you will find yourself frustrated. You begin to think s/he is just being lazy, or maybe being passive aggressive. Observe and learn. At each stage of the disease, agitation and confusion make your person less able to perform tasks once done well. Persons with Alzheimer's disease/dementia (AD) will have good days and bad days. Get used to observing your person's behavior and level of understanding when you are asking him/her to do multiple tasks. This will guide you in developing healthful support ideas.

Can follow a verbal list of several tasks
Losing the ability to do sequential tasks is part of the disease.

For instance, when you lay a shirt out on the bed, you have an expectation that your person will put it on, button it up and tuck it in. Start to gain

Notes:

some understanding of the complex nature of these kinds of requests and how to simplify them as time goes on.

• Observe your person's behavior and level of understanding when you are asking him/her to do multiple tasks.

• Get used to giving simple direct requests. This will minimize confusion and frustration even in early stages of AD

Can follow visual/written instructions for tasks
While your person can still look at a written list and follow the instructions, this is an invaluable tool.

• If your person has used "to-do" lists before, make lists of tasks, this will be a familiar, non-embarrassing way of gently reminding him/her of things to be done.

• If there is anything that is forgotten on a regular basis, put a sign up. "Flush the toilet."

• Use post-it notes and lists in early stages as reminders for appointments and tasks.

Can do 2 or 3 tasks one after another
Get used to creating routine tasks. When things happen in the same order at the same time of day, it is easier for your person to adapt and s/he will experience less confusion.

• Give simple 2-3 step directions. Limit the directions to your person's abilities.

• Order tasks such as tying shoes, putting on jacket and putting on hat so that they become routine.

• Lay out items in a similar order each time to promote what memory your person can bring to the circumstance.

• Observe routine and try to repeat activities in the same order.

Can only do one task at a time
As the disease progresses, the ability to follow progressive instructions diminishes.

• Only ask the person to do one thing at a time.

• Continue to lay out clothing and items to be used so that the pattern of use looks the same.

• Make your request close to the time that you want it done. Retention at this stage will be very poor.

Must be supervised for the whole task
At this stage your person will need clear routines and direct one-step prompts.

• Patiently supervise your person.

• Give him/her simple one-step directions.

• Limit your difficulties by having your person work closely with you performing a simple repetitive task. The simplicity will give your person comfort and keep him/her safely occupied. This works well when working with household, gardening and craft/art activities.

Cannot understand instructions
Be patient with your person. It will seem at times as if you are speaking a foreign language.

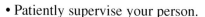

Notes:

• Patiently supervise your person.

• If s/he can still follow through with the rest of a task after being jump-started, then allow him/her to finish even if s/he is slow.

• If directing fails, then perform the task along with your person. Allow him/her to do as much of the task as s/he can.

• At some point you will have do the task for your person , but postpone this as long as possible so that your person does not lose what little skill s/he may still have.

TIME

General

One of the initial diagnostic tests for determining Alzheimer's disease/dementia (AD) is the ability to tell time. Clinicians will tell the person a time then ask them to draw the hands on a blank clock face. Time can be fairly easily lost as a mathematical ability. Yet the concept of time may linger on as well as the ability to use a watch. Many people find a great deal of comfort in wearing a watch even if they never look at it. It is as if you can capture time by simply having it on. Time can lend a structure to the day. If your person cares a great deal about time; or if time was involved in one of his/her professional skills, then it may last longer. S/he may have more brain pathways concerning this subject or ability.

Can tell time

If your person can still tell time whether it is on a traditional clock or a digital one, encourage this skill. You might want to be careful with expensive watches even at this early stage. You never know when your person might misplace it. Thinking a bit ahead is the safest path.

• Encourage the patient to wear a watch (if possible not an expensive one).

• Ask the time even if you can just glance at a clock.

• Make a schedule and regularly do things at a certain time of day.

• Ask your person to remind you of the time.

Can't tell time but enjoys watches and clocks

In our modern culture where time is integral to our lives; losing ones time orientation can be distressing and agitating. Being old enough to tell time and getting your first watch marks a milestone of responsibility and competence in many people's lives. Managing time is one of the skills of a responsible person. When your person starts to lose that ability, it can be very difficult for him/her. S/he may want to conceal the fact out of embarrassment and may have been compensating for some time. It's important to recognize how important a symbol time and the symbols of time are to your person. In the early stages of AD, people notice their losses. They begin to notice when they consistently fail to do the things they once did. They still remember how sharp they were and they mourn the loss. A person with AD with time displacement will not know how long s/he has been at a place. You may go to dinner at a relative's and your person wants to leave as soon as s/he gets there.

If time is important to your person, try these things:

• Take your person out and shop for a watch (with a large face and big numbers).

Notes:

• Get a fun watch, like a Mickey Mouse watch.

• See if your person can still read a digital clock. If your person can read the numbers but not understand the significance, s/he may still feel competent.

• Try and do some things at the same time each day and remind him/her when it's time. (if this brings them comfort and structure).

• If you visit someone and your person begins to obsess about time, keep him/her engaged and occupied in some simple distracting activity.

• Keep extra inexpensive watches around the house, so that those that are misplaced can easily be replaced.

• Let your person wear a watch even if he or she cannot tell time. It can be a comfort and a symbol of competence.

Time is Irrelevant

If your person doesn't care about time, it may just be their temperament, resignation to their loss, or the progression of the disease. When your person begins to lose track of time s/he may switch day for night and wander the house at night. This is called "sundowning." This can be disruptive for the family. Check the section on **Sleep** (page 128) for ideas that deal with "sundowning" behaviors.

POSSESSIONS

General

This is such a big issue in Alzheimer's disease/dementia (AD), which is why it is a separate category. When a person with AD loses or gives away a possession, s/he may forget and then blame others for the loss or theft of the item. This often causes great interpersonal stress. This problem can reoccur day after day and be quite wearing on the caregiver and family.

Can handle and organize personal possessions

Even the most competent of us has a hard time with organizing our things. If this skill is still intact in your person, put it to good use.

• If your person had good organizational skills s/he may enjoy a system of post-it reminders to remember various daily activities, tasks or appointments.

• Encourage your person to begin to pare down his/her possessions or store them away, so that there is not so much to keep tack of. This will keep organizational skills functioning longer.

• Begin to limit the volume of things that your person is responsible for.

• Ask to keep valuable things that your person may own. You never know at what stage your person will begin pulling valuables out of a drawer or box to have a look at them and then stash them in some easily forgotten hiding place. This is especially important in dealing with precious gems and small items. I found my mother-in-law's diamond ring in a vase four years after she died.

Hoards possessions

Hoarding is a common behavior in AD. It can include never throwing out the trash, retrieving things from the trash, saving packaging from items, taking items from other places like day care, and stashing items from all

over the house. It does not matter who owns the item; a person with dementia will usually not consider that fact, just that the item may be needed now or in the future. My mother-in-law used to consistently take paintbrushes from the day care. Scolding did not work. I just quietly removed and returned them. We did let her keep some so she wouldn't take them all. Being non-confrontational and understanding can elicit more cooperation than being accusatory.

- Just observe the behavior, if you are so fortunate as to see it. Check the hiding place later, and return the item to the right place when your person is not watching.

- Remember scolding will not change the behavior. It may just scare, annoy, or provoke a catastrophic emotional reaction. (The emotional centers of the brain are heavily affected by AD making them more prone to over reacting to negative situations.).

- If you observe your person taking a very valuable item and you don't want to lose track of it, make up an excuse to thank them for finding the object. Tell them how grateful you are and return it to its rightful place or a new safer place.

- If you are missing an item and did not see it taken, you may just have to wait until it turns up.

- You can try to observe your person's favorite hiding places and routinely check them for missing objects. Be careful that your person does not see you going through his/her possessions; this may set off a wave of paranoia or negative emotional reactions. If you are caught, don't act sheepish, but make up an excuse that you dropped something in the room. Even enlist your person's help in looking for your "lost" item. Be creative in these kinds of circumstances. You will make up your own tricks and ways of handling this problem.

Loses possessions

It is normal in the course of the disease for people with dementia to begin to misplace things. They put things down and can't remember where they put them. Or they may hide things and forget where they are hidden. You are unlikely to be successful by asking your person where s/he put the item and it may only agitate them.

- Help your person look for the item.

- If your person is becoming obsessed about finding the item, distract him/her with some food or another activity. You can tell him/her you will look for it later. This can buy you time to produce a substitute if you have one.

- Get spares of eyeglasses, hearing aid batteries, anything small that might be misplaced.

- If the house is less cluttered, it is easier to find things that have been misplaced.

- Limit the number of hiding places by locking rooms or closets that are not needed.

- Check wastebaskets before emptying them.

- In the winter, always use the same jacket and hat for your person

- Put mittens or gloves on little elastic suspenders hooked to the arms of the jacket.

- If you can, stuff the person's hat into the arm of his/her jacket.

Notes:

Loses possessions and blames others

As stated above, a person with AD will often misplace things. Women may continually wonder about where their purse is. Men may have a constant worry about their wallet or keys. Understand that your person lives in a confused and often baffling world; blaming others for the loss of items is a way of bringing some order to their disappearance. Unfortunately, it can make other people feel annoyed or angry.

• Try the same ideas as in the previous section plus the following:

• Take valuables such as silver, or rings and put them in your own safe place. When the person mentions that they have been stolen assure him/her that you have it safely stored. If your person insists on seeing it, get it but don't show your hiding place. Be firm about returning it to the safe place.

• Buy a couple of purses or wallets.

• Put together several sets of keys.

Doesn't care about possessions

This behavior certainly represents less turmoil for caregivers, but may indicate a state of major loss for your person. If this occurs earlier in the disease, have your person screened for depression.

REMINISCENCE

General

One interesting thing scientists have found is that, while persons with Alzheimer's disease/dementia (AD) tend to lose their short-term memory in the early and middle stages, their long-term memories tend to endure. Your Uncle Joe may forget that he just took a bite of pudding, but may remember in detail what was harvested from the family garden in the fall of 1942.

In the mid 1960's Dr. Robert Butler developed a technique called "life review" that encourages elderly persons to recall their past in detail. It is a technique that brings to the mind of an older person: this is who I am, that is what I've done, this is what I leave behind and this is what I feel about it. The technique, though usually practiced with intact elderly people, seeks to assist people to regain or reinforce a sense of identity, to cope with the stresses of aging, and helps to build self-esteem. When practiced with persons with AD, it allows the person to skip their recently lost memories and mine the older memories that recall a time when they were full of life, vigor and ability. Often, persons with dementia leave a session of reminiscence feeling valued, happy and satisfied.

Loves/Likes to review the past

If your person is into past memories, this is an excellent time to get him/her to tell his/her life story. You can be record these stories for your family.

• As opportunities present themselves, ask your person about his/her past.

• Be patient and listen.

• Listen to where memories take your person; don't worry about the information being provided.

• Be empathetic and confirming. "I'll bet that was hard for you."

• Focus on what your person was good at. Allow him/her to share skills and knowledge.

• Be sensitive to your person's expressions and body language.

• Encourage the joyful memories and don't dwell on the painful ones.

• Validate expressions of anger or grief. "It sounds like people treated you pretty badly over that incident. That must have been hard."

• Keep emotions flowing and encourage your person's success stories. "That must have been a great time for you."

• Old movies, pictures, sounds, and even smells like popcorn popping in the kitchen, cider mulling on the stove, can trigger memories. Keep the senses engaged.

• Savor the moments; the reminiscences are a journey not a destination. Your person spends a great deal of his/her life living in real time. These memories though referring to past events are very real for him/her and s/he can make better sense of these than what happens day to day.

• Record stories for posterity

• Create a picture book of the memories with the input of your person. Once created be careful to preserve it. If shared with your person unsupervised in later stages, s/he may forget the reason it was made or who the people are in the pictures and try to destroy it. The same is true for single photos or other mementos.

Indifferent about the past

It is useful for you to remember that the object of active remembering is to stimulate still active brain cells, to keep the person engaged and happy, and to relieve some of your stress. As cold as it sounds, at this point there is no recovery and no therapeutic path back to health for the person with dementia. The disease continues to progress and the person continues to regress. Your person may be able to maintain what is there, but s/he won't recover what is lost. There is however the issue of moments of "lucidity" where there are sudden bursts of knowledge or concern. Follow these moments with your person; you may learn some interesting bit of history before the thoughts fade again.

There will come a time, as the disease progresses, when even the distant memories will fade. Scrapbooks you may have put together for remembrances may stimulate memories; but as the memories fade, the person's enthusiasm will likewise fade.

• Gather what history you can when you can, then put it aside.

• Concentrate instead on helping your person remember a few small things in the present: where the bathroom is, where his/her coat is kept, and where to sit when s/he eats.

• To the extent possible within an active family's life, try and keep some schedule with your person. Eat at the same time. Even if it's just one meal a day that is on a schedule.

• Have your person go to bed at the same time. This structure is similar to what is needed for young children.

Destroys objects from the past

This is one of the disturbing aspects of **retrogenesis** – reverse development. The Alzheimer's disease/dementia (AD) regression mimics in reverse, the development of a child. Persons with AD often reach a stage much like

Notes:

a two or three year old where they destroy things. It is often objects that relate to his/her past. This is a stage where the person needs consistent supervision, especially with valuables and mementos.

- Essentially, you will need to "childproof" the person's environment.

- Remove valuable or memory-rich objects from their immediate environment.

- Put pictures and scrapbooks away or supervise their use.

PHYSICAL
EYE SIGHT

General
It is quite common for any elderly person to experience some decline of sight in varying degrees. As the dementia progresses, the person with Alzheimer's disease/dementia (AD) begins to interpret sensory perceptions differently than prior to the disease. Visual loss is not a symptom of dementia. Maintaining a person's visual acuity may help stave off hallucinations and delusions. Ask your person regularly to read or report what s/he sees, to figure out if there are changes in eye sight that might need correction.

Can see without difficulty
- Get vision checked regularly; especially if there is any question about what your person is seeing.

- Encourage reading and other visual activities. This can help to keep the brain stimulated and the eyes healthy.

Must use glasses
If your person has used glasses for much of his/her life, there are some built-in habits. If your person only uses glasses for reading, s/he may have more problems keeping track of the glasses.

- Avoid contact lenses, unless your person is in very early stages of AD and has ingrained contact lens habits.

- Get glasses with plastic lenses.

- Get glasses frames that don't break easily.

- Get a spare pair of glasses. If they're just magnifying glasses, get several drugstore pair to have on hand.

- Keep the prescription handy in a safe place where you will remember where it is.

- Inquire frequently as to how your person is seeing.

- Have a regular place for the glasses when your person takes them off to bathe or sleep.

- Use neck cords that keep glasses handy to help avoid loss. (Glasses neck cords may present a safety hazard. Just be aware of how your person handles the glasses and cords.).

Complains about vision in spite of glasses
This could be an indication of another physical condition or it could just be the brain of the person with dementia starting to misinterpret visual stimuli.

- Get your person a vision exam with an ophthalmologist.

• Report the symptoms to the regular physician.

Has difficulty with vision tests

Eventually, your person will have difficulty reporting to the doctor what s/he sees in the vision test.

> • Even if you don't suspect s/he is having this trouble, get into the habit of discretely explaining to the doctor or examiner your person's condition.

> • Accompany your person into the testing room to help interpret his/her responses.

> • Due to the difficulty sorting symptoms, it would be better to use an eye doctor/ ophthalmologist to perform the eye exams verses a commercial provider.

Has cataracts

Cataracts cause blurry and dimmed vision as well as complicating how your person interprets what s/he sees. The blurry vision can lead to a worsening of hallucinations or delusion. It will also mean that you will need to watch this person much more closely. The possibility of hurting him/herself is increased especially if s/he doesn't have the cognitive ability to compensate for the handicap. Reaching for something on a hot stove can be catastrophic. If the disease affects depth perception, your person can begin to knock things over increasing the danger of burns and other accidents. Cataract surgery can be an option.

> • Supervision is very important at this point.

> • Have cataracts evaluated by an eye doctor/ophthalmologist and discuss treatment options.

Has cataracts that need surgery

It depends on the health of your person and the degree to which the cataracts affect his/her quality of life. When the cataracts "ripen" a person's vision will be very limited. Work with your physician. The recuperative time may well complicate the whole process.

> • Consult with your MD on the pros and cons of surgery. Your person's general health will be a factor.

> • Surgery and hospital stays disorient persons with Alzheimer's disease/ dementia (AD); this needs to be a factor in the decision.

> • Out patient surgery, if safe, is preferable.

> • Eyesight is a valuable sense. Blurred vision is disorienting and may aggravate AD.

> • There is special bandaging over the eye after surgery. Your person may try to remove it unless supervised.

> • The whole episode will be taxing for both of you but may still be worth the effort.

Legally blind

If your person has been legally blind for a long time then s/he have made adaptations for the condition and you will be able to adapt care suggestions. If this condition is new, then you have a very difficult situation of trying to have a person make adaptations with a brain that is neither quick to adapt nor understand. This fortunately is not a frequent occurrence. The National Center for the Blind can be of help.

Notes:

• Alzheimer's disease/dementia does not make people blind, but in late stages when your person has regressed to an infant stage, s/he may not be able to understand what s/he sees. At this stage s/he will not recognize people and may only acknowledge happy faces the way a child does.

HEARING

General

Difficulty with any one of the five senses can lead to increased difficulties with interpreting sensory data correctly. If your person has no hearing difficulty, continue to check and monitor his/her hearing health. Test it informally by calling their name softly while their attention is not directed to you. If s/he responds, it is a quick and easy way of monitoring. If there are any questions about your person's ability to hear, you should consult your MD. Do this sooner rather than later in case the problem is treatable.

Has no hearing difficulty

This is where you want your person to be and to stay. Hearing loss is not a symptom of dementia, and would most likely come from some other health condition. It is important to have an MD evaluate any losses of hearing

• It is normal for older people to experience mild diminished hearing.

• One of the first signs of hearing loss is a difficulty in understanding words. This symptom may mimic sensory interpretive losses caused by Alzheimer's disease/dementia (AD).

• Your person may also have difficulty hearing high frequency sounds such as doorbells or some telephone ringers.

• Have symptoms checked out by an MD so that you can separate dementia related problems from those caused by other diseases/health conditions.

Complains of hearing difficulties

Hearing loss can take two forms. In Conductive hearing loss the sound fails to reach the inner ear. A person will hear him or herself fine but can't hear others. The other type is Sensory Neural hearing. Here, the inner ear is affected and the person will speak loudly in order to hear himself.

If your person complains of hearing difficulties, have his/her hearing checked. It is painless and catching something early can make a big difference. Hearing loss can be caused by a myriad of problems, infections, trauma, dental problems, extended exposure to loud noises, a build-up of earwax and even diabetes. Diminished or lost hearing may be an early indicator of some other health condition. It is good to become aware of any physical condition that may account for behaviors. Move on these things early because as the disease progresses, the ability of the person with AD to report changes or increased loss will diminish.

• Try to limit the amount of background noise during conversation.

• Turn off the radio or TV.

• Don't whisper around your person- include them in conversations.

• Modulate your voice into their hearing range, but don't shout.

• Consult an MD for advice on the need for hearing devices.

Hearing impaired

If your person requires a hearing aid, do some research and get a sense of what is available and what will be the least invasive and the most helpful for your person.

- It may be useful to get a spare hearing aid for him/her. There is a great potential for loss of the device.

- You also might take charge of putting the unit in and out as part of the routine.

- Your person may forget why the device is needed and be resistant to its use. Be gentle. Don't get angry and force it on him/her. You may let them live in his/her quiet world and only use the hearing aid for important occasions.

- You may have to develop alternate forms of communications, such as hand/body signals.

- Picture cards can be helpful.

Deaf

A total loss of hearing can be very disorienting to a person with Alzheimer's disease/dementia (AD) , especially if s/he has to perform a task. Otherwise it may be quite peaceful. Consult the National Organization for the Deaf for help and support.

- Be sure all hearing loss has been evaluated by an MD

- Learn alternate ways of communicating, such as hand/body signals and picture cards.

SMELLING

General

It's a fact of life that as we age our sensory perceptions begin to dim. This is true of sight, hearing, and the sense of taste and smell. AD also seems to cause a diminished sense of smell. It is not considered a definitive symptom but may be a precursor. Smell is closely linked to taste. This may contribute to a decrease in appetite.

Can smell without difficulty

At this point not much needs to be done to preserve this state. Any sense should be utilized and enjoyed while it is still intact. The senses create stimulation for the brain, keeping it active and possibly creating new pathways.

- If your person likes diverse tasting foods, then offer them often.

- Enjoy and comment on the aromas of flowers and pleasant natural surroundings.

- Utilize aroma therapy (pleasing or calming scents). Try lavender or sweet orange.

Reports things do not smell as good as they used to

The sense of smell has begun to diminish at this point. Take it in stride and try to stimulate what is left.

- Inquire as to what continues to smell good.

- Provide doses of whatever your person still enjoys.

- Test new things from time to time just to see if they are pleasing.

Notes:

Notes:

• It is good to provide continued stimulation to encourage the smell-taste connection.

Uses too much perfume or cologne

Clearly things can go awry as the senses begin to decline.

• Encourage your person to use it only after s/he has bathed and the senses are refreshed

• Water down the perfume or cologne with water

• Take control of the perfume or cologne and dispense it for him/her.

Diminished appetite

This is the result of continued loss of the sense of smell as well as other factors affecting the health of your person. Maintaining an appetite allows the person to continue to consume adequate nutrients.

• Try foods that have a little savorier flavor like soups, stews and meats with gravy.

• Add a bit of soy sauce or other savory sauce or even sweetener to add more flavors to vegetables.

• Try using flavor enhancers such as Mrs. Dash, bacon bits, even catsup and mustard.

• Identify flavors that are still pleasing and use those more often.

• If your person's favorite food is nutritious, then offer it frequently.

• Consider vitamin supplementation.

• See section on **small appetite** (page 54)

Seems unresponsive to smell

There is the unfortunate possibility that smell can become very diminished in dementia. The loss of the ability to smell will diminish the quality of life, but more than that it may prevent adequate nutrition.

• Consider vitamin therapy.

• If your person objects to swallowing pills, consider a powdered or liquid preparation that can be added to juice. See **Nutrition** (page 55)

TALKING

General

The early stages of Alzheimer's disease/dementia (AD) are often the most difficult for family members, friends, and your person. By its nature, the disease has taken hold and begins to show symptoms before it is diagnosed. Everyone is dealing with the odd collection of symptoms. In the early stages your person may be very aware of these losses. Speech remains intact in early stages, but will slowly erode into word retrieval problems (where the person cannot find the right word or phrase to describe a person or circumstance). We all fumble a bit as we enter middle age but the loss with dementia is more dramatic.

Speaks with little difficulty

This of course is ideal but as with all the senses, it is good to stimulate speaking abilities.

• Engage in conversations that interest your person.

Notes:

• Ask him/her to recount stories about his/her life.

• Encourage conversations with other people, even if you are afraid your person may make a mistake.

• Encourage social situations such as day care, which will stimulate conversation.

• Talk with your person about his/her day especially if s/he has been away from you in day care or on some other respite experience.

Uses words incorrectly, or cannot find the right word

Using words incorrectly, mixing the order of words or using inappropriate words can be frustrating for a caregiver because it means that your person's communications skills are starting to deteriorate. It can also be very distressing for your person as well, especially if s/he is proud of the ability to articulate or speak well.

• Practice being direct and literal.

• Don't make word jokes that require mental acuity and can be misunderstood.

• Be specific about what you want the person to do. "Go to the closet and get your coat and put it on." Rather than "Go get ready."

• Gently support your person in finding the words s/he may be searching for. Be matter of fact and clearly not demeaning.

Difficulty making sentences and expressing ideas

The loss of communication abilities can be incremental and slow. Your person may be inarticulate in the morning and lucid in the afternoon. S/he may go for days as if in a fog, then reappear as if quite fine. This can be heartbreaking. It's almost as if your person has recovered, and then two days later s/he is in a fog or doing repetitive behaviors.

Keep a steady hand. Be grateful for the good days but don't come to rely on them. When your person appears to be recovering, it is just his/her brain trying new routing. Unfortunately this too, will likely eventually deteriorate.

• Talk with your person about real concrete things not abstract ideas. Lack of understanding can make him/her feel "stupid".

• Use the specific names for things and people. Avoid general pronouns, it, he, she etc.

• Look at your person when you talk; make sure s/he is paying attention and indicating understanding.

• Nod and affirm that you understand what your person is saying.

• If your person is having trouble finding the word – gently provide it, don't make him/her stretch. For example, your person might gesture the motion of putting something on and say. "I'm looking for the things I put on my feet." – "Your socks? Your shoes?" and hold them up to see.

• Keep your statements and questions short.

• Reduce background noise. Turn down a radio or TV.

• Stop working and give them your full attention.

• Remember you are not doing recovery therapy. Be accepting, gentle, kind and giving.

Notes:

Makes little or no sense when speaking

Don't assume that your person doesn't understand what you are saying. Remember the retrogenesis model. If you reference child development, children understand long before they can speak. In a person with AD, the ability to speak is diminished before the ability to understand language is lost. There still may be strange periods of understanding called "lucidity." During these times persons with dementia may regain some of their former skills and be able to suddenly speak about something they care about—only to lose it again.

- Persons with dementia retain the ability to understand emotional context.

- Try and learn what your person is saying with his/her rearranged vocabulary context. Subtly interpret for your person in social situations.

- Don't carry on a conversation about your person with a third party as if s/he is not in the room.

- Be patient.

- Be direct when speaking to your person. Either give your full attention or tell him/her that you don't have time right now. Don't make him/her guess.

- In tense situations, remain calm and directive. Avoid yelling at your person.

- Ask or say one thing at a time. Don't try and carry out complex commands or give your person too much to digest at once.

- Keep a simple conversation thread going: "Do you like the pie?" "What kind of pie did you eat as a girl?" "Did you make pies?"

This will keep your person engaged and require little of him/her but to follow your thought thread.

Nonverbal and uses gestures to talk

This may not happen until your person is generally not very responsive. It is worth mentioning because every person' losses are individual. Many supporting activities you would use for a small child may be helpful and comforting at this stage.

- Be reassuring with your nonverbal communication (gestures, facial expressions and looks). Your person will remain sensitive to these after the ability to talk has been diminished.

- Look at your person when you talk. Interpret verbally his/her nods and gestures to indicate you understand. "You're cold aren't you? I'll get you a sweater".

- Smile and search out his/her reaction. "You like that, don't you?"

- Hold your person's hand when you want him/her to concentrate on what you are saying.

- Use a soothing and calming voice when your person is agitated.

- Pacing or hitting objects are signals of your person's frustration.

- Even if the situation is tense, attempt to remain calm and even.

- Read to your person even when s/he cannot talk much. Just like small children the spoken word /story can bring comfort.

Cannot speak and does not respond to conversation

This is a very late stage loss, and your person may have a many other skills that are lost.

Notes:

• If you touch your person, do so in their line of sight. This will prevent him/her from being startled or becoming frightened.

• Hold your person.

• Pet his/her hand.

• Rock with your person.

• Play soothing music.

• Be gently directive and still explain what is going on when you are working directly with your person.

• Be patient.

• Continue read to your person even when s/he cannot talk much. Just like small children the spoken word /story can bring comfort.

Randomly shouts or cries out

During the later stages of the disease, repetitious muttering, shouting, moaning, nonsense sounds or other agitated sounds are common.

• Try to discern any meaning behind repetitive sounds. Some may express discomfort or hunger.

• Be patient and try to calm your person.

• Give your person some attention.

• Use distraction.

• Use some of the suggestions in the previous "Cannot speak" section.

TELEPHONE

General

The telephone is both a device and a symbol of communications. As a device, we take it for granted, using it every day to keep in touch. For a person with dementia a telephone can be a powerful symbol of contact with others as well as a symbol of his/her competency. When this ability starts to unravel, using the phone can be an extremely confusing task.

Can use telephone

If this ability is still intact, then encourage your person to answer and make calls.

• If you have a portable phone, this is a good time to attach some kind of finding device (available through specialty stores and catalogues – some home portable phones now have paging devices built in). We all have trouble putting the phone down somewhere only to have it covered by other household items. Since you do not know when the casual loss of the phone may change to purposely putting it in a place where it may be very difficult to find, it is good to plan ahead and decrease the frustration in your life.

• Encourage friends and family to call now while your person can enjoy the conversations.

• Develop a system of writing down phone messages. This will help preserve abilities especially as the tasks start to become more difficult.

• If eyesight is a problem, buy a phone with bigger numbers on it.

• Figure out which phones you would like to use and then stick with them. Change will be harder to handle in the future.

Notes:

• Monitor your person's use of the phone, so that you can anticipate new strategies that you may need to use. Talk to family and friends who call, and encourage them to let you know how your person is managing phone conversations.

Can take messages

If your person can still take messages let him/her do so, but monitor this activity carefully.

• Be aware of telephone solicitors. They can take advantage of demented people. Make sure your person doesn't have easy access to credit cards.

• Monitor your telephone bill to make sure s/he isn't doing any business or making long distance calls that you are unaware of.

• Remind your person that it is okay to politely hang up if a conversation gets too confusing.

• Let friends and family know that your person may hang up if s/he becomes confused. This will decrease misunderstandings and hurt feelings. People can call back at another time.

Cannot take messages

Your person may still be able to use the phone to make or listen to calls but may not be able to accomplish the complex task of message taking. If your person cannot take messages, the telephone can cause him/her a lot of anxiety.

• Let your person know that you don't expect him/her to take messages.

• Do not scold or berate your person if s/he tries to take a message. Your person may not realize s/he has done anything wrong or may be defensive and afraid of your reaction. Just thank him/her for the efforts and go on.

• Get an answering device or Voice Mail.

• Turn off the phone when you are gone and not able to supervise the phone.

• When you are there, keep the portable with you.

• Use call-forwarding and forward incoming calls to your cell phone or office.

Cannot dial or hang up phone

Your person may still be able to talk on the phone, but at this stage parts of the task are unraveling.

• As with small children, you will have to supervise phone calls.

• Limit phone calls to people with whom your person is comfortable.

• Dial the phone for your person and hand him or her the receiver and return it when s/he is finished.

Cannot use the phone at all

By this time your person may not understand what the phone is or how to handle it. It is also possible that a few parts of the function may still be intact and your person may still be curious about the phone. Like a child, s/he may want to investigate or play with the phone.

• Reassure your person that he/she is not responsible for answering the phone.

• As with small children, you will have to supervise the phone.

• Let family and friends that would make calls know what is going on.

• By now you are probably not leaving your person unsupervised, so let your sitter know what your person can or cannot do.

• If your person is still anxious to play or work with a phone, buy an extra one and don't hook it up. Leave it out where s/he can see it and play with it.

• If the phone begins to create anxiety, remove it from sight and hopefully your person will begin to forget all about it.

EATING

General

Eating is one of the most critical areas of life from the aspects of nutrition to the sensory pleasures. Eating involves many different aspects of a person's life. It is a pleasure activity. It is a social activity. It involves motor skills to manipulate knives and forks. It is about appetite and favorite foods. In general you will have to take charge of mealtimes and what is eaten. In the beginning you can involve your person with meal preparation, cooking, and clean up. This will help preserve his/her skills and abilities as well as provide a sense of purpose, usefulness, and involvement in family activities. Be gentle with your person, s/he will make mistakes and this is no longer a learning situation but one of preserving skills.

Plastic mats, perhaps a smock or adult fashioned bib and napkins nearby can help quickly prevent accidents or clean up after a messy eater. Don't rush the meal.

Has no difficulty eating

For as long as you can, keep your person involved in the mealtime activities and gathering. It will give him/her an emotional lift, encourage more inter-actions with the family, and keep the person interested in eating and staying at the table.

• Get meals into a routine. This may be hard if you have children and a busy life but it will decrease confusion in your person.

• Try not to have your person eat alone. This encourages erratic eating habits and poor nutrition. The social aspect of eating is very important.

• Develop a good idea of what your person does and doesn't like. Cater to those likes and dislikes when nutritional and possible.

• Give your person tasks at each meal. This will keep him/her involved and feeling helpful and valuable.

• Try your person out with a variety of different tasks until you find a few that s/he is good at and assign those routinely.

• See the section on instruction as your person begins to have trouble following directions.

• Have your person sit in the same place each meal

• Be forgiving when mistakes are made. Your person is doing the best s/he can and this is not really a learning situation but a preserving one.

• Observe for changes in your person's eating patterns, so that you can make nutritional changes as needed

Notes:

• If your person snacks or overeats, put food out of reach or put more nutritional items in the front of the refrigerator.

• When eating out, avoid dark fussy places. Your person will make mistakes in eating and etiquette. Your attitude and expectations need to be adjusted before you go.

• If your person has no memory of eating the last meal and wants to eat again soon, keep some celery, or carrot sticks or other nutritional but low calorie nibbles around. Also try getting them to drink some liquid, like water or juice to fill their stomach.

• Be careful with food prepared in the microwave, it can come out much too hot to eat and your person may lack the judgment to refrain from eating it.

• If your person has diabetes, special care must be taken. Your person will not remember what is permitted and what is not. You will have to be in charge of food choices and consult with a nutritionist as needed.

• Don't worry about messiness; just take steps to help cleanup.

Has difficulty using utensils

At some stage, knives and forks may become a problem for your person to use. If s/he seems to stop eating for no obvious reason, s/he may have forgotten how to get the food to his/her mouth.

• Use gentle verbal reminders like "pick up your fork" or place the fork in the correct position in his/her hand. The memory of the process may just kick in and s/he will continue eating without difficulty.

• Eat with your person and encourage mimicking of your maneuvers. Get his/her attention and casually demonstrate picking up food with a fork and putting it in your mouth. Try not to make it a lesson but more of a gentle game.

• Introduce utensils with large built-up handles.

• Use only spoons. They are easier to use than forks and are less dangerous.

• Use special plates with curved lips to make food easier to scoop out. (These plates and utensils may be obtained from the children's section of a department store, or possibly from a local medical supply store or on-line at dementia products web site – See product appendix for details.)

• Cut up food into easy bites that do not require using a knife. At this point it may be best to do this before the food is brought to the table. Then your person will only see himself as competent and not being overtly treated like a child.

• Order foods at restaurants that you know your person can eat without difficulty or embarrassment and will not require cutting at the table.

• Try finger foods such as sandwiches, chicken, pizza, hot dogs, and fresh fruit like diced apples or bananas.

Hides food in napkin or hoards food.

This is not an unusual behavior in dementia. Your person may be hiding food in a napkin because s/he doesn't want to eat it, or think that they are "saving" it for later. Remember **retrogenesis** – reverse development. This behavior is very child-like.

• Eat with your person and observe for this behavior. You will most likely catch it when you get your person up from the table.

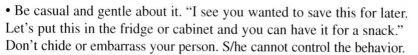

Notes:

• Be casual and gentle about it. "I see you wanted to save this for later. Let's put this in the fridge or cabinet and you can have it for a snack." Don't chide or embarrass your person. S/he cannot control the behavior.

• Try serving one food at a time so your person doesn't feel overwhelmed with the amount of food.

• Offer a number of small meals verses three large ones, but do keep your person coming to the table to eat with you or the family.

Tries to feed pets from plate

This is nurturing behavior but does your person no good nutritionally and may cause problems with pet behavior.

• Clear the pets from the eating area, so that your person is not distracted from eating or tempted to feed pets. This will help keep your person eating what s/he needs.

Needs assistance with eating food

At this stage, your person's motor skills are beginning to fade. In addition to needing help cutting food, you can also anticipate spills both on the table and on the person. Your person is now beginning to regress to a child-like state in the ability to handle eating. Eating will get messier and messier. Your person may smear food on her/himself or on the table just like a one-to-two-year-old. You will want to keep your person independent for as long as possible, but you will have to set the limit and decide when to begin spoon-feeding.

• Anticipate spills, have plenty of napkins and paper towels available.

• Place a mat below your person's chair to catch the residue.

• Use plastic tablecloths or mats for easy clean up.

• Use high-rimmed plates, un-tippable cups, suction cup plates, double handled cups as needed (these items can be ordered from catalogs, obtained from the children's section of a department store, or possibly from a local medical supply store or on-line at a dementia products web site – See product appendix for details.)

• Use bibs/smocks to contain spills. Adult bibs can be sewn fairly easily and can be fashioned to seem more utilitarian or fun and less demeaning. You can even wear one yourself to make your person feel more comfortable.

• Offer straws with drinks to minimize spills.

• Cut up food into easy bites that do not require using a knife. At this point it may be best to do this before the food is brought to the table. Then your person will only see himself as competent and not being overtly treated like a child.

• Order foods at restaurants that you know your person can eat without difficulty or embarrassment and will not require cutting at the table.

• Try finger foods such as sandwiches, chicken, pizza, hot dogs, and fresh fruit like diced apples or bananas.

• Be aware of the temperature of the food. Make sure foods are not too cold and never too hot.

• Test hot liquids on your wrist as you might a baby bottle, before trying to offer them to your person.

Notes:

Has difficulty chewing or swallowing

Persons with Alzheimer's disease/dementia (AD) will slowly lose the ability to remember to chew and eventually the motor ability to chew. This again is another difficult stage in **retrogenesis – reverse development.**

• Be sure dentures fit properly

• Remind your person to chew with every bite. This will help keep him/her from trying to swallow without properly chewing the food.

• Serve softer foods like eggs, meat loaf, soft meat, that require little chewing

• Grind up food to likeness of baby food.

• Serve some baby food, but spice it up a bit.

• Use vitamins and nutritional supplements to boost nutrition lost as the volumes of food eaten grows smaller.

• In later stages choking can be a serious problem. Ask your MD for a speech pathologist (SLP) evaluation. These skilled therapists not only help people with speech problems after accidents or strokes; they also work with feeding and swallowing difficulties. These SLP services are available in a clinic setting or in the home if your person is home bound.

Unable to feed self

If your person refuses all food s/he may have regressed to child/infant stages of development and have unlearned how to eat. Continue to support any ability to self -feed, but most likely you will be transitioning into hand feeding.

• Begin hand feeding with regular table food cut into bite size bits for easy chewing and swallowing.

• If s/he has any difficulty with bite size foods move to soft blended foods that are easily swallowed. Soups, whipped potatoes, smoothies (blended fruit drinks), creamed vegetables, and applesauce are other soft foods that can be used.

• You may need to consult a nutritionist at this time to determine amounts and kinds of foods that are needed.

• You can use commercial baby foods and pudding with protein supplement, as alternatives if you are tired of preparing soft food. A nutritionist will have other helpful suggestions when you are too tired to cook.

• Hand feeding someone three meals a day can be taxing. This is a good time to ask your support network for help.

• Be aware of the temperature of the food. Make sure foods are not too cold and never too hot.

• Test hot liquids on your wrist as you might a baby bottle, before trying to offer them to your person

• In later stages, choking can be a serious problem. Ask your MD for a speech pathologist (SLP) evaluation. These skilled therapists not only help people with speech problems after accidents or strokes; they also work with feeding and swallowing difficulties. These SLP services are available in a clinic setting or in the home if your person is home bound.

• Some of the suggestions in the next section will also be helpful.

Needs to be fed in bed

If you are feeding your person in bed, you will need to have a few preparations. You need to keep your person's head up so that there is less chance of gagging and choking.

- Allow your person to participate in the process in any small way s/he can. Even if it is just holding something for you.

- Get them in a comfortable sitting position or as close to sitting as possible.

- Prop/elevate your person's head

- Cover the area with protective, washable covering.

- Use straws and keep the sipping end above the level of the liquid to avoid air bubbles.

- Support your person's head while drinking.

- Pureed foods can be drawn through a large drinking straw.

- Be aware of the temperature of the food. Make sure foods are not too cold and never too hot.

- Test hot liquids on your wrist as you might a baby bottle, before trying to offer them to your person.

- Fill the spoon partially full and bring it to your person's lower lip placing a bit of pressure there to encourage him/her to open the mouth. Then put the food in and give him/her time to swallow.

- Talk calmly and slowly to your person offering them encouragement.

- Give yourself time to feed your person.

- Don't rush or s/he may pick up on your mood and become agitated or eat poorly.

- Observe and record the best eating times for your person and which foods are tolerated.

Cannot Swallow

At this point your person has regressed to the infant/fetal stage of development and has to be totally maintained with artificial nutrition usually through a nasogastric tube. This demands 24 hr. care and you will need help to keep your person in your home. Many people place their loved ones in a facility at this point because the care is so demanding. As your person sinks deeper into the disease s/he may be essentially in a vegetative state. This is a time of hard decisions and you will need support from your family and network. Read the section on **End-of-life decisions** (page 217).

- When the tube is placed, someone at the MD office, hospital, or home care will instruct you on the use and maintenance of an NG (nasogastric) tube. If the use of the tube is suggested and no one talks about education or support, bring it up. Ask questions.

- This is not a difficult procedure to learn but you will need support and proper instruction.

APPETITE

General

Dementia will eventually have an affect on appetite. In the beginning things may be going well but it is good to be prepared for future difficulties. Changes can range

Notes:

Notes:

from out of control consumption due to damage to the appestat mechanisms in the brain to picky eating and obsession with certain limited selections. In later stages your person will lose the mechanical ability to eat and appetite will be negatively affected as well. Loss of the ability to smell may also be a factor. Either way some conflict over food is likely to occur.

Has good appetite

If your person has a good appetite, try and preserve this as long as possible. Meals and mealtime are central to social interaction and maintaining good nutrition. Food preparation and kitchen chores help preserve competencies in your person.

- Serve foods your person likes, especially if the foods are nutritious.

- Let your person talk or express feelings and thoughts about food. This can stimulate the desire for food.

- Involve your person in the food preparation (Refer to cooking and meal prep section for ideas).

- Involve your person in serving of the meal.

Has large appetite

This can be a good or bad thing depending on your person's basic metabolism. If the metabolism is high coupled with a large appetite and an appropriate weight, then this will not be much of a problem. It could be a good factor in maintaining nutrition. If this is leading to weight gain, and you feel it is unhealthy, or is going to make life difficult, start with techniques that could work with any overweight person.

- If this is not causing any weight gain or obsession problems, then leave it alone.

- Your person's physician should be consulted to determine his/her healthy weight.

- Scale back your person's portions, and remove extra food from sight.

- Restrict sweets.

- Increase vegetable and fruit offerings, especially as snacks.

- Think about vitamin supplements to keep good levels of nutrition.

- Work gently with your person. If he/she needs an unhealthy snack from time to time to maintain happiness, then let him/her have the snack. Life at this point is relatively short and some pleasure is important.

- Monitor weight once a week or so to see that there is no trend toward excessive weight gain.

Has small appetite

Issues around a small appetite will depend on the weight of your person. Your physician should be consulted to determine the healthy weight of your person. If s/he is overweight then a smaller appetite could be a good thing. If your person is underweight then there may be a need to increase caloric intake to provide nutrients for the body and the brain. The trick is to use nutrient dense foods because your person will not likely increase the volume that s/he is willing to eat. If your person's appetite is such that it is causing weight loss or if there is a sudden unintended trend to lose weight, notify the your doctor. Some weight loss is common in the later states of Alzheimer's disease/dementia (AD), but it may also signal the onset of another disease or dental problems.

Notes:

• Make sure your person is not experiencing any obvious dental or physical discomfort that may affect his/her appetite.

• Make sure there are no mechanical problems such as the inability to use a knife or fork. See section on eating for tips working with this problem.

• Sit and eat with your person. This may make the time more comfortable and social.

• GENTLY -Coax them to eat. Reminders can be important because your person may simply have forgotten to be hungry.

• Offer a number of small meals/snacks verses three large meals.

• Remember that over-exerting yourself to force your person to eat things s/he does not like or to increase the amount of food that is eaten can be defeating for both of you. Do the best you can and do not torture yourself. SNEAK IN THE NUTRIENTS.

• Offer their favorite foods, even milk shakes and ice cream, which contain calories and nutrients. Put an extra egg and or banana in a milk shake. Put fruits and/or ground nuts on ice cream.

• Offer calorie dense snacks like peanut butter on crackers or celery (depending on your person's teeth and ability to chew).

• Add wheat germ to hamburgers or sprinkle on salads.

• Grind up nuts and put them on cereal or salads.

• Balance the meal between protein and carbohydrates. Don't offer extra bread if your person is not consuming meat and vegetables. White bread has calories but very little nutrient value. Switch to denser wheat bread if your person will eat it.

• Offer fresh squeezed or pressed fruit and vegetable juices.

• Add vitamin rich protein powder to milk shakes or pudding to increase nutrient value.

• Blend extra steamed vegetables and mix into commercial or homemade spaghetti sauce.

• Keep food cut into small bite size portions so that they are more easily eaten

• Use dietary supplements-like Ensure.

• Monitor weight once a week or so to make sure there is no major trend toward weight loss.

Has no appetite

Loss of all appetite can be common in the later stages of AD. Continue with the same ideas in the section on poor appetite. If your person refuses all food, s/he may have regressed to child/infant stages of development and have unlearned how to eat. If s/he will eat if hand fed, then you will have to spoon-feed soft blended foods that are easily swallowed. For more ideas, refer to **Unable to feed self** (page 52).

NUTRITION

General

What we eat is so important in how we feel and how much energy we have. As you age, your metabolism tends to slow (unless you are a high level exerciser), your needs for calories will generally decline, but your needs

Notes:

for vitamins and nutrients will remain the same or may increase. You will have to get your nutrients from less food, if you do not want a weight problem.

When a person has Alzheimer's disease/dementia (AD), nutrition becomes a key factor in preserving function and preventing any mental declines that can be caused by malnutrition. In early stages there may be some opportunity to make some changes in food habits, but as the disease progresses the potential is more variable. Food habits are very individual and affected by culture, ethnicity, and the regions of the world that you have lived in. What ever your normal cuisine is, it is best to build from that. Luckily preparing and eating more nutritious meals benefits everyone in the household. There are several basic ways to eating more nutritious meals:

• Eat a balance of meats and vegetables (fresh or frozen are better than canned) and complex carbohydrates (grains and breads with higher fiber and wheat content).

• Try to avoid pre-packaged foods. They are often very high in starchy carbohydrates that are not very nutritious (despite what the label says). If cooking just doesn't work for you, try to find the quick meals that advertise nutrition verses size.

• Simple meals are fine. Serving a chicken breast with steamed vegetables and whole-wheat bread is a good nutritious meal. Serving more food does not make it more nutritious. You have to work the meals around the reality of your life.

• Keep fats at a moderate level. New studies are showing that it may be the combination of fats and high sugars in the bloodstream that encourage clogged blood vessels. Low fat diets in general have only made people fatter. Foods that say they are low fat often increase the sugar content to keep the product tasting good. Eating large quantities of low fat processed foods can increase your sugar consumption and thus increase weight and health problems.

• Decrease the amount of sugar in your diet. This holds for underweight and overweight people. Sugar used to be touted as an energy booster and still is a major ingredient in many foods. Over time it has been established that sugar may give you a spike in energy; but soon after, it also causes a great drop. Now experts talk about foods that have low glycemic response, which means that they may be sweet but don't push your blood sugar up so high and don't give you the matching plunge that can make you feel shaky or hungry. This yo-yo effect over the years can predispose some people to diabetes. The bottom line is that sugar has no nutrient value.

• Use fresh fruit or sugar-free products as desserts. Fruit offers vitamins and minerals, but the sugar-free products may just offer taste and comfort (not to be entirely overlooked). Due to the popularity of low carbohydrate diets there are many more sugar-free products on the market. If your person is under-weight then you may want to use regular ice cream to increase calories.

• Try to increase fiber in your diet through grains, fruits, and vegetables.

• Drink plenty of water. It is vital to all systems in your body, and dehydration can cause a number of health problems. Observe your person's liquid consumption and keep it at an adequate to optimal level (6 to 8 glasses of water/liquid per day). Don't fully count caffeinated beverages,

Notes:

as they have a tendency to dehydrate.

• NOTE: Having stated the ideal above, in dementia anything can happen. Some food is better than no food. You will need to keep nutrition in mind when selecting what you and your person eat, but only you will know what foods will be tolerated. This is not a good time to make good nutrition a religion, but just use it as a supportive strategy to keep health and function up.

Vitamins

Vitamin therapy is an important and growing area of health promotion and disease prevention. It is unfortunate, but many of our vegetables no longer have the same levels of nutrients that they used to have (largely blamed on soil depletion). A number of experts now feel that it is difficult to get optimal levels of nutrients through food alone.

Vitamin therapy ranges from simple daily multiple vitamins to complex combinations of individual nutrients. It takes quite a bit of time to become knowledgeable about the myriad of nutritional products you will see in a grocery store or a health food store. Good daily vitamin supplements can make you feel better. Wade in slowly because the cost can add up.

• Start with a simple daily multiple vitamin supplement for you and your person. You can get advice from nutrition-oriented professionals, friends, or health food store staff.

• Try a few different kinds and see which kind you both tolerate the best. Compressed pills are harder to digest than capsules, but it generally takes fewer of them to make a daily dose. Check to see how many pills it takes to get the full daily dose. It can range widely from 1 to 2-3 at each meal. If there are a number of pills, you can work up to the full dose over time

• Swallowing on demand can be a problem for your person. Obtain vitamins in liquid or powder form (You can take apart capsules of powder as well.) and mix them with juice in the morning. Usually this does not alter the taste much.

• You can read up on vitamins and minerals and learn their specific applications.

• Calcium supplementation is something to consider and may be necessary if you or your person have a tendency toward osteoporosis or thinning of the bones. It is good to get a test for this (bone density scan) so check with your MD. Osteoporosis is a major factor in bone fractures in the elderly.

• Vitamin C is an excellent supplement to add to your daily vitamins (Most combination pills have a basic dose). Vitamin C is water-soluble and you can take fairly large doses (1000 mg +) without harm to the body. It can help in building immunity and reducing the effects of colds and sinus problems. Check for powder forms if swallowing is a problem.

• Check with your MD about the possibility of B-12 deficiency, which can cause worsening Alzheimer's disease/dementia (AD) or similar symptoms. This is usually done as part of differential diagnosis to rule out other causes of AD. If you are not sure if it any testing was done ask your MD. B-12 deficiencies are usually treated by monthly injections (You can learn to do this).

Notes:

• If you want to engage in large-dose vitamin therapy for AD or other conditions, consult an alternative physician, naturopath, or nutritionist to help you, as it can get very complex. Your person may not be able to swallow or digestively tolerate large dosing and you will need some one to give you advice on the effects and managing the doses. If you want to see what has been suggested there are few popular books on nutrition that list vitamins, minerals, and suggested dosing for AD (Such as Dr. Atkins vita-nutrient solution).

• A daily protein and vitamin-powdered supplement can also be healthful and easy to consume. These can be used as a meal substitute if you are on the run. They can be mixed with water, juice or milk. Be careful to read the carbohydrate levels on these products. The popular pre-mixed ones sold in the grocery stores often have huge amounts of sugar. Unless you or your person is substantially underweight it does not do you much good to consume your nutrients with large amounts of sugar.

Eats well-balanced meals

If you and your person eat nutritious balanced meals already, you are way ahead of most people. Keep up the good work. The challenges for you may be in keeping your person eating this way. It is hard to say whether his/her appetite will increase or decrease as the disease progresses through the early to middle stages. In later stages it is likely that s/he will eat less and less until the ability to chew and swallow is lost. Through all this there will still be feeding difficulties.

• Keep meals simple, offering one type of each nutrient (proteins, carbohydrates, vegetables, and fruits or other desserts). This keeps choice and confusion to a minimum

Eats an unbalanced diet

Do the best you can to turn the situation around. Your person's willingness to eat new foods is individual. Some people become more docile and will eat anything placed in front of them. Others will be picky and only eat what is familiar.

• Make a list of all his/her favorite foods as well as those that are just tolerated.

• Make a comparison list with more nutritious alternatives.

• Slowly work between the two lists and make whatever strides you can.

• Read sections on encouraging various nutrients.

• Sneak in nutrients where you can See section on **Small appetite** (page 54).

• Monitor weight once a week or so to make sure there is no major trend toward unwanted weight loss or gain.

PROTEINS

General

Protein foods include meats, fish, eggs, milk, cheeses, beans, soy, cereals, whole grains, and nuts. Proteins are needed to repair and preserve body tissues, especially muscles. They also support the formation of antibodies to fight infection. Meat proteins are high in iron. Milk and related products

Notes:

supply calcium and vitamin D. The need for proteins does not change as you age, but can vary depending on your person's health. Illnesses or health problems can create an increased need for protein. These foods can be expensive and more difficult to prepare and swallow. Thus some elderly people eat less than they should. Two to three servings are needed per day.

Likes to eat protein foods
If your person enjoys eating a variety of proteins, continue to rotate favorites.

• Keep offerings to one-to-two choices per meal. Offering a wide range at one meal may confuse your person and may cause them to eat too little. Obviously holidays are different and should not be obsessed over, as their value is as social as it is nutritional. One or two days won't make a difference.

• Rotate favorites so that your person does not lose the concept of variety. You do not want to encourage getting picky and only wanting to eat one or two kinds of protein.

• Don't worry if your person becomes picky anyway. Many changes are unavoidable.

Protein foods enjoyed:
List here for those who may be offering respite for you.

• MEATS:

• FISH:

• CHEESES:

• MILK AND MILK PRODUCTS:

• BEANS AND LEGUMES:

• CEREALS AND GRAINS:

Has difficulty eating protein foods
This can be due to appetite, the ability to smell, the ability to chew; or s/he may just dislike protein foods.

• Make sure that there is not a mechanical problem like chewing. If there is, then cut or grind the protein into edible bites. Make sure all the food is prepared in a chew-friendly way so that your person will be able to eat all that is offered.

• Create combinations of foods such as vegetables or grains mixed with meat or cheese.

• Eat with your person and let him/her see you eat and enjoy your food. This will encourage imitation that may be able to override dislike of the food.

• Sneak protein into the food through blended combinations like protein powder milk shakes. Add protein powder to other milk products if s/he can tolerate them.

Notes:

• If your person does not tolerate milk or milk products, protein drinks can be obtained that are soy based. Soy ice creams can also be used to spark up the flavor. Soy products can be found in health food stores and even in some large grocery stores.

• Add protein powder to grains or in creamy sauces.

• Puree the meat to a baby food consistency. It may be more palatable.

• If purees don't work you may have to resort to having them drink protein beverages either made yourself or commercially prepared. These beverages can be purchased in grocery and health food stores

CARBOHYDRATES

General

Carbohydrate foods include bread, pasta, grains, vegetables, potatoes, fruits, sugars, and fiber. Carbohydrates are a source of general energy as well as for the growth and maintenance of body cells and the central nervous system. Complex carbohydrates provide added nutrients as well as fiber (whole grains, and whole grain products, vegetables, and certain fruits) to prevent constipation and promote a healthy colon. Simple carbohydrates like white flour; rice, bread, pasta and sugar contain fewer nutrients and are found in most processed foods. These lower nutrient foods are everywhere in our grocery stores.

Carbohydrates are usually inexpensive, easier to prepare, and chew. Thus some elderly people eat more of these foods and less protein. Whole grain breads, cereals, and pastas should be used if possible. If white flour and rice are part of your cultural diet, then it is probably not going to be appropriate to change them unless your MD has requested the change for health reasons. Cakes, pies, desserts, and junk food also fall under the simple carbohydrate category and should be minimized because they have almost no nutritional value. If you prefer a high protein diet then you can get sufficient carbohydrates from vegetables and lower sugar fruits without having to consume breads and pasta. Four or more complex carbohydrate servings per day are recommended

Likes to eat complex carbohydrates

This is a great place for your person to be. S/he will be a healthier person. However long your person has been building this habit, it is important to keep it up.

• Continue to offer a variety of favorites on a rotating basis.

• Keep offerings to one-to-two choices per meal. Offering a wide range at one meal may confuse your person and may cause them to eat too little. Obviously holidays are different and should not be obsessed over, as their value is as social as it is nutritional. One or two days won't make a difference.

• Fresh fruits and vegetables are preferable to canned.

• Keep fruits or other dessert until the end of the meal. Sweetness is often a taste sensation not affected by Alzheimer's disease/dementia (AD). Even a good eater may be attracted by sweet foods and forego needed proteins and vegetables.

• Don't worry if your person becomes somewhat picky anyway. Many changes are unavoidable

Complex carbohydrates enjoyed:

List here for those who may be offering respite for you.

Notes:

- BREAD:

- PASTA:

- GRAINS:

- VEGETABLES:

- POTATOES:

- FRESH OR DRIED FRUITS:

Likes mostly simple carbohydrates

Many people are used to higher quantities of simple carbohydrates like white flour, rice, bread, pasta and sugar that are everywhere in our grocery stores and in most processed foods. These foods are low in nutrition but easy to buy and to eat. Even people who eat well balanced diets often have simple carbohydrates that they love. The trick is to keep consumption within reason and not to exclude more nutritious foods. If your person is underweight, you will have to carefully examine what portions of precious calories are consumed in simple carbohydrates. If it is high, then your person will be in a nutritional compromise and you may have to make some changes. If your person is overweight, then this may be a cause and any changes could be very valuable.

- Make a list of all his/her favorite simple carbohydrate foods.

- Make a comparison list with more nutritious alternatives

- Slowly work between the two lists and make whatever strides you can.

- Your person may adjust to eating what is put in front of him/her because s/he may forget what else is out there.

- Offer small fixed portions of these foods and do not offer more until more nutritious foods are consumed.

- Offer nuts instead of chips. You can buy low to unsalted nuts if sodium is a problem.

- Offer small portions of more nutritious carbohydrates and follow with one of your person's favorite simple carbohydrates as a reward.

- Offer fresh fruit raw or cooked as an alternative to high sugar and high fat products.

- Fresh squeezed fruit juice may be an alternative to high fat desserts.

- Mix higher nutrient foods in with simple carbohydrates- such as hamburger and vegetables combined with white rice. You may even be able to transition to brown rice if this is an enjoyable combination. Try to make these combinations yourself verses buying processed versions.

- If you have to resort to pureed foods due to chewing or swallowing problems, you can mix in some whole grains to add nutrients and fiber.

Notes:

Simple carbohydrates enjoyed:
Record your person's allowable favorites here for anyone helping you with respite.

- BREAD:

- PASTA:

- RICE:

- VEGETABLES:

- POTATOES:

- FRUITS:

- DESSERTS:

- SNACKS:

Has a passion for sweet foods
This is a hard one to tackle if your person wants only sweet foods verses any nutritious foods.

- Try making things sweet with sugar substitutes like Splenda. It has no after taste and you can cook with it. Other non-sugar sweeteners lose much of their sweetness in cooking or have an after taste. Stevia- Plus is a sweet herb and health food product. It is probably better for the body, but the taste may take some getting used to.

- Combine low sugar jams with peanut butter for sandwiches.

- Make peanut butter banana sandwiches. They are both nutritious and sweet.

- Add protein powder to puddings and milk shakes to provide needed nutrition.

Eats very few carbohydrates
This is a more unusual situation than over-consumption. If your person prefers a high protein diet then s/he can get sufficient carbohydrates from vegetables and lower sugar fruits without having to consume breads and pasta.

FATS

General
Fats are a part of a balanced diet, but should be consumed in moderation. If your person does not consume many simple sugars, the fats s/he consumes will probably not be a problem. If your person loves sweets then s/he is probably consuming excess fats. Hydrogenated fats are considered very unhealthy and can be found in certain margarines, processed foods, and commercial desserts. Saturated fats are found mostly in meat

Notes:

fats. Polyunsaturated fats are in many of the vegetable oils on the market. Monounsaturated fats are considered the best fats, and a main source is olive oil.

Likes to eat fats

This would mostly pertain to meat eaters, butter or margarine consumers, and those who love desserts.

- Fats combined with sugars are the unhealthiest.

- Trim your meats to decrease over-consumption of saturated fats. If your person is underweight, this is less of a problem.

- Use butter or margarine that is without, or low in saturated fats.

Fats allowed:

Record your person's allowable fats here for anyone helping you with respite.

- FATS:

- SPREADS:

- OTHER:

Eats very few fatty foods

This is not much of a problem, unless your person is underweight. If so, then mix healthy fats into his/her foods and this help will increase calorie consumption.

VEGETABLES

General

Vegetables are a great source of nutrients. They are low in fat and calories. Increased portions of vegetables in the diet can decrease weight and increase fiber.

Likes to eat vegetables

This is wonderful for a balanced diet. If your person consumes more vegetables s/he may likely eat fewer simple carbohydrates.

- Continue to offer a variety of favorites on a rotating basis.

- Keep offerings to one-to-two choices per meal. Offering a wide range at one meal may confuse your person and may cause them to eat too little. Obviously holidays are different and should not be obsessed over, as their value is as social as it is nutritional. One or two days won't make a difference.

- Fresh vegetables raw or steamed are preferable to canned. Frozen organic vegetables are now more readily available.

- Don't worry if your person becomes somewhat picky anyway. Many changes are unavoidable.

Vegetables enjoyed:

Record your person's favorites here for anyone helping you with respite.

- VEGETABLES:

Notes:

Vegetables hated:
Record your person's dislikes here for anyone helping you with respite.

> • VEGETABLES:

Refuses to eat most vegetables
This is unfortunate because vegetables are so nutritious.

> • Make a list of the vegetables that your person will tolerate.

> • Offer these on a routine basis.

> • Cover nutritious vegetables with sauces and see if your person will eat them.

> • Grind up vegetables and put into a tomato sauce and serve with pasta, rice, or potatoes.

> • Sneak in nutrients where you can See section on **Small appetite** (page 54).

FRUITS

General
Fresh fruits are a great source of nutrients and fiber. They are low in fat and calories. Fresh fruits are preferable to fruit juice, which contributes mostly sugar to the diet.

Likes to eat fruits
This is wonderful for a balanced diet. If your person consumes more fresh fruits, s/he may likely eat fewer simple carbohydrates.

> • Continue to offer a variety of favorites on a rotating basis.

> • Offer fruits at the end of the meal so that they do not compete with needed protein.

> • Keep offerings to one-to-two choices per meal. Offering a wide range at one meal may confuse your person and may cause them to eat too little. Obviously holidays are different and should not be obsessed over, as their value is as social as it is nutritional. One or two days won't make a difference.

> • Fresh fruits are preferable to canned.

> • Don't worry if your person becomes somewhat picky anyway. Many changes are unavoidable.

Fruits enjoyed:
Record your person's favorites here for anyone helping you with respite.

> • FRUITS:

Fruits hated:
Record your person's dislikes here for anyone helping you with respite.

> • FRUITS:

Refuses to eat most fruits

This is not too much of a problem because nutrients can be made up with vegetables.

- Make a list of the fruits that your person will tolerate.

- Offer these on a periodic basis to encourage variety in the diet.

- Sneak in nutrients where you can (See section on low appetite).

FLUIDS

General

Fluid consumption is very important for body function, elimination of waste, and electrolyte balance. As we age, there is a tendency to have a diminished sense of thirst; and with Alzheimer's disease/dementia (AD), thirst may just be forgotten. Dehydration can cause serious problems for a person with AD. Be careful of hot beverages, your person may not handle them safely.

Likes to drink water

This is a wonderful habit.

- Offer 6-8 or more glasses per day.

- If your city water is highly chlorinated or contaminated, you may want to consider commercial bottled water. Chlorine is just another factor that does not cause disease but does not promote health.

Likes to drink coffee

The caffeine in coffee is a stimulant but also tends to dehydrate the body.

- If there are no other problems and your person really enjoys coffee, continue to let him/her have it in moderation.

- Be observant of your person's behavior around coffee. If s/he becomes agitated or exhibits behavior problems, you may want to change to decaffeinated. Coffee is also a social beverage so keeping it in some form or other may be a comfort for your person.

- Offer increased water to offset the dehydrating effects of coffee.

Likes to drink fruit juice

Fruit juice is less desirable than fresh fruit; but if your person likes it, keep it in the diet.

- Buy the best quality juice you can buy. Health food stores are usually the best source.

- Buy fresh squeezed or pressed juices as they come into season in your area.

- Buy a juicer and make your own fresh juices. This can be an expensive, but it could be well worth it if your person likes vegetable juice as well.

Doesn't like to drink fruit juice

If your person does not like fruit juice, it is not much of a problem. Fruit juices contain mostly sugar and offer few nutrients unless they are fresh squeezed or locally pressed in season.

- Use vegetable juice if added fluids are needed.

- Try offering the best quality juice from a health food store. This is only useful if you need to get additional fluids in your person.

Notes:

Notes:

Doesn't like to drink water

This can be a problem because water is necessary for life and health.

- Offer flavored waters, sparkling or not. The stores are full of different kinds.

- Offer sugar-free lemonade or other flavors.

- Offer fruit juice in moderation because it may bring excess sugar and be too filling.

- Offer sodas or other favorite beverages in moderation.

ALCOHOL

General

Alcohol consumption can be a distinct problem in Alzheimer's disease/ dementia (AD). Moderate consumption of alcohol is seen as offering some health benefits in healthy adults; but for persons with AD, even small amounts can cause reduction in his/her ability to function. If your person has never been much of an alcohol consumer most of his/her life, a sip now and then in a social situation probably wont hurt them. If your person has been a chronic consumer or is possibly an alcoholic, then you may have to wean down and/or eliminate his/her consumption. Continuing to consume large quantities of alcohol can affect appetite causing the threats of malnutrition or dehydration. AD has a powerful effect on the Limbic system causing emotional swings. Alcohol can compound this and your person may act irritable, resentful, stubborn, or even hostile. This creates behavior management problems and functional dependencies for you to have to deal with. If you are deeply concerned that your person is an alcoholic, talk to your MD before making any changes.

NOTE: Alcohol can cause side effects or reactions if your person is also taking prescription drugs. Check all labels carefully and consult with your local pharmacist or an M.D.

Does not drink alcoholic beverages

This is best for all the above reasons. Alcohol compromises function.

- Offer simulated non-alcoholic beverages that look similar to what everyone else may be drinking (like sparkling cider). This can promote social comfort if that is an issue.

Likes to drink alcoholic beverages and still wants to drink occasionally

This is the category for those who have consumed small to moderate amounts for some time in their lives. It is wise to supervise all consumption of alcohol because each person will react differently. As AD progresses your person may become more sensitive to alcohol and lose function quicker. Note the issues on prescription drugs stated above.

- Offer alcohol in small glasses to sip.

- Offer small amounts during social occasions, perhaps just enough to give the mixed drink a light flavor of alcohol.

- Wine can be put into a spritzer to dilute the alcohol.

- Limit consumption to social occasions.

- Offer simulated non-alcoholic beverages that look similar to what everyone else may be drinking. This can promote social comfort.

Has a chronic drinking problem

This is a great challenge for a caregiver. You have both the chronic drinking and the declines from AD to be concerned about. As stated above, chronic alcohol use can magnify the expected losses from AD. You may not be able to deal with this problem on your own. Seek help from your person's physician to develop a plan for weaning off of alcohol. It is possible that after the weaning your person may begin to forget the desire for alcohol, which is a blessing.

• Stay away from saying things that emotionally charge the situation, such as placing blame or trying to elicit guilt. Your person needs to retain his/her dignity and self-esteem.

• Offer a drink at the usual time. Try a smaller amount and dilute successive drinks each day

• Offer small amounts of the desired alcohol only at mealtime.

• Try to distract your person if s/he wants more or offer a diluted drink as a second drink

• Try different glasses so that the amount doesn't look like less.

• Wine can be put into a spritzer to dilute the alcohol.

• Limit consumption to social occasions.

• Offer simulated non-alcoholic beverages that look similar to what everyone else may be drinking. This can promote social comfort.

• Lock up all alcoholic beverages out of sight. Keep the key in a hidden place. This will help your person forget how to find them and how to get to the alcohol.

• Don't unlock the storage area and pour drinks in front of your person. This will reinforce his/her memories of how to get into the cabinet when you are not looking.

• Move all the alcohol to a new place, or remove it from the house.

• Don't drink in front of your person. This will also reinforce drinking memories and behavior.

• Report any problems with behaviors or function, possibly caused by alcohol to your MD and seek help.

SMOKING

General

There is no doubt that smoking is bad for your health. Unfortunately it is a habit that is relatively easy to pick up and is very addicting. If you can convince your person to stop smoking, this will probably enable more of his/her brain cells to last longer and you will avoid future difficulty when your person cannot manage cigarettes safely.

Does not currently smoke but is ex-smoker

If your person does not currently smoke, be thankful. If your person is an ex-smoker, don't let him/her start up again. Keep any signs of smoking out of the house just in case your person may have forgotten that s/he quit. If you smoke, keep exposure of your person to the second-hand smoke minimal. Studies have shown that second-hand smoke is very damaging to the lungs of others. You want to keep your person's health at optimum levels if possible.

Notes:

Notes:

Enjoys smoking and can manage it

Besides the health risk smoking poses, with persons with Alzheimer's disease/dementia (AD), it will inevitably become dangerous. Recognize however that smoking is a powerful addiction and the smoker, whether demented or not will say and do nearly anything to avoid withdrawal. Smoking is a vasoconstrictor and may accelerate the progress of the AD.

- Observe your person while s/he is smoking so that there is no danger of fire.

- Try to convince your person that quitting smoking now while s/he is still functional is the best thing for health and brain function.

- Slowly try to wean your person off of cigarettes. Cut down a bit each day to slowly wean the body from the need for nicotine.

- Consult your MD about the possible use of the nicotine patch to cut down cravings.

- Smoking will seriously compromise any medications that are taken to promote better brain function.

Cannot hold or dispose of cigarette or cigar properly

This is a dangerous situation, and could lead to a fire that could hurt your person and others in the house. At this point, s/he will have to go to you to get the cigarettes; so there are a number of things you can do. If your person reacts with anger or threats to any deprivation of cigarettes consult with your physician, there may be some adjustment/addition of medication that will help.

- Do not let your person smoke alone.

- Watch your person closely when they smoke and provide him/her with a large, easily seen ashtray to keep stray ashes to a minimum.

- Continue to try to wean your person off of cigarettes; one less a day may helpful.

- Take charge of cigarettes, cigars and matches. Give them out only when asked.

- Keep the paraphernalia out of sight, so your person can forget about it. This is where bad memory can work in your favor.

- Consult your MD about the possible use of the nicotine patch or gum to cut down cravings.

- Try distraction, your person may be able to temporarily forget that s/he even wanted a cigarette. Offer a cookie or some other desirable treat that may give oral pleasure. Offer another activity such as a walk, or other recreational fun.

Does not want to quit smoking

If your person is obstinate about continuing to smoke after you have tried a number of the above suggestions, have your MD talk to him/her, a physician may be taken more seriously.

- Remember the power of distraction. Offer food or another activity that may take your person's mind of the need for a smoke.

- Consult your MD about the possible use of the nicotine patch or gum to cut down cravings.

Family members smoke

If you or other family members smoke, it will be very hard to get your person to quit.

• Try to keep your person away from second hand smoke and any smoking paraphernalia. Remember, out of sight—out of mind.

• Keep all smoking activity outside the house.

Has lung disease

If your person already has lung disease from smoking or any other source, it is important to stop smoking and keep second hand smoke away.

• If your person is on oxygen, then it is imperative that smoking ceases for safety as well as health.

• Consult with your MD about creating a smoking cessation program using "out of sight-out of mind" ideas as well as nicotine patches and gum.

ARMS AND HANDS

General

Motor use of various parts of the body is key to everyday function. Motor skills deteriorate in dementia but usually at a slower rate than the mental and emotional areas. Depending on the condition of health of your person, these problems may or may not be important. Each person is individual and any other diseases or conditions they may have will make a difference in the level of loss. Exercise and use are key elements of preserving motor function in any area.

Able to grasp and carry things

As with all intact skills, they need to be exercised. If your person seems reluctant to do something or shows some weakness, encourage activity during the day. Always check with your MD if there are any sudden physical losses. These could be a result of underlying conditions such as a stroke. Motor skills begin to deteriorate as the disease progresses.

• Exercises that engage hand-eye coordination will help preserve function.

• Exercises that improve the arm and hand strength will help preserve function.

• For women, keep purses a manageable size and weight.

• For men, give them packages that have handles or plastic bags with handholds. A bag held in the arms, like a grocery bag can slip.

• Continue to let your person open cans and jars if s/he can, especially if s/he is helping in the kitchen and the situation is safe.

• Have your person play simple ball games with grandchildren or neighbors.

Has weak grasp but can hold or carry things

Be alert for other conditions that may exaggerate this condition. Elderly persons often have arthritis or problems with muscles or joints. Be gentle with them.

• Have your person evaluated for other conditions. Perhaps medication can help.

• For women, lighten the purse.

• For men, keep packages light.

• Try to remember to avoid handing your person things they may drop like a glass of water.

Notes:

• When you do hand your person something, make sure s/he has a sure grip on the object.

• Open jars and cans for your person to minimize frustration.

• Watch your person around doors and make sure s/he can negotiate them. If not, keep vital ones open and close off areas where you don't want them to go.

• Weakness in this area can also affect the ability to get dressed and general hygiene. Consult **Dressing self** (page103) **and personal hygiene** (page 110) areas of the book for more supportive ideas.

• Consult with your MD if hand/arm weakness is of concern. A consultation with an occupational therapist (OT) could be helpful to preserve function.

Cannot hold or carry things/ has little use of hands and arms

If your person gets to the point where s/he can't hold or carry things, but is still alert then have him/her examined by your physician (unless you already know of a condition that explains his/her losses-like arthritis). This loss should only happen in late stage Alzheimer's disease/dementia (AD) when your person is getting generally unresponsive and probably bed-bound.

WALKING

General

Walking is a very important skill for function and recreation. Inability to walk presents a myriad of problems in caring for your person. You will want to encourage walking on a regular basis. It will help preserve function and improve general health. Walking can have a calming affect on mood and improve sleeping.

No difficulty walking

For your person, walking is an activity that can afford them a great deal of pleasure and exercise. As the disease progresses, they may begin to shuffle.

• Take routine walks with your person. It will improve your health as well as the health of your person.

• Take walks at the local zoo, park, or nature sanctuary. These walks can bring both of you some peaceful time. Try to leave your worries at home and relax with your person.

• As your person becomes more childlike s/he may re-experience the natural environment with new eyes and renewed pleasure.

• Encourage your person's interest in nature. It may bring a soothing spiritual connectedness with the earth and its rhythms.

• Have your person walk with you when you are shopping. This can be frustrating if s/he has a tendency to wander off while you are deciding about purchases. If possible, keep your person engaged finding food items for you. This may keep him/her lingering in the aisle with you.

Has difficulty walking

If your person is having difficulty walking outside, it may be because of his/her medication, the progression of the disease or just the advancement of age. Check with your physician if any of these changes are sudden.

• If the problem is mild still keep up your walking program, just slow down a bit to accommodate.

• If your person develops a tendency to stumble or fall, have him/her evaluated by your MD. This would be a good time for an evaluation for osteoporosis if it has not already been done.

• There can be a big danger in even small falls. In a person with osteoporosis or thinning of the bones, a small fall may cause a hip fracture. Your person can have an alteration in his/her experience of pain and may not express it clearly. A high tolerance for pain can mask actual damage that may need to be treated. Check with your physician if you have any doubts.

• Ask your physician for a physical therapy (PT) evaluation in an outpatient setting, so that you can understand what the problem is and what to do about it.

• Provide input as to your person's routine and habits to the physical therapist, so that s/he may be able to create a workable exercise plan for your person.

• Follow the physical therapy plan as outlined. Get back in touch with the physical therapist if you have questions.

• The therapist can also evaluate for equipment needs such as canes that can help increase function and steadiness.

Shuffles

Though it may be normal for a person with Alzheimer's disease/dementia (AD) to develop a shuffle as the disease progresses, have him/her checked by your physician. Changes in gait could also be the result of a stroke or a medication s/he is on.

• If the problem is mild still keep up your walking program, just slow down a bit to accommodate.

• If your person develops a tendency to stumble or fall, have him/her evaluated by your MD. This would be a good time for an evaluation for osteoporosis if it has not already been done.

Stumbles

Stumbling but not falling means that your person is able to regain some balance after losing it. Unfortunately stumbling usually can lead to falling and the dangers of bone damage. Prevention is very important at this stage.

• Have your person take your arm when walking.

• Make sure your person is getting sufficient exercise. This will maintain muscle tone.

• Have railings installed in hallways and bathrooms.

• Follow from behind and support your person with their belt. This may prevent you from falling if your person falls.

• Stumbling may be accompanied by other failures of motor skills. Be alert for your person's inability to match his/her zippers or hold things properly, or if your person misses his/her plate with his/her fork, etc.

• Clear the area of obstacles. Take up slippery throw rugs. Take up pots and vases and decorative objects that might present a hazard.

• Put safety tape on steps to highlight them.

Notes:

• Ask your MD for a physical therapy evaluation, just in case there are some other underlying causes that can be re-mediated. A physical therapist (PT) can also do home safety evaluations to recommend changes in the home that you might have missed.

Tendency to fall

It is often difficult for a person with dementia to regain balance after a stumble and s/he may often fall. If your person shows a tendency to fall, have them checked out by a physician. It could be medication, delirium or even a small stroke, and not just the progression of the disease.

• Clear obstacles from their usual pathways.

• Make sure all pathways in the house are well lit.

• Keep door to the basement locked.

• Be there to observe and help your person around any stairs you may have in your house.

• Block off the stairs with a childproof gate, keeping your person on the ground floor with you while you are working. (Note that using a gate is a suggestion. Your person may still get through it. Keep trying until you find something that works. Gates can be found in children's departments in large stores.

• Make sure that relatives or friends don't rearrange things in an effort to streamline. This could confuse your person and make stumbling and falling more likely. A foot cushion or coffee table placed where it is unexpected may provide a stumbling block. An end table or easy chair, to your person, may be a handhold or leaning post that's not there when they need it.

• Install railings especially in the bathroom.

• Put non-slip material on the bottom of the bathtub.

• Ask your MD for a physical therapy (PT) evaluation, just in case there are some other underlying causes that can be re-mediated. A PT can also do home safety evaluations to recommend changes in the home that you might have missed.

Difficulty walking up or down stairs

Even if your person does not have obvious problems with stumbling or falling, the stairs are a problem waiting to happen. If you live on a single floor, it will be much easier for you to manage.

• Offer your person an arm to support him/her going up or down the stairs (especially public stairs with wide distances between the banister and the wall.

• If you have problems keeping yourself steady on stairs, ask others to help. Most people don't mind being asked to help an elderly person on the stairs. If one says no, ask someone else.

• Make sure banisters and railings in your house are secure.

• Block off the stairs with a childproof gate, keeping your person on the same floor with you while you are working. (Note that using a gate is a suggestion. Your person may still get through it. Keep trying until you find something that works) Gates can be found in children's departments in large stores.

• Install barriers to keep him/her from either going up or down stairs. Make sure it is high enough so s/he does not trip.

Notes:

Needs to use cane

Hopefully you will have had a consultation with a physical therapist (PT) to fit the cane. The therapist is responsible for training you and your person in the proper use of the cane.

- If you have not had a consultation with a physical therapist, ask your MD for one.

- Your person should practice with the cane on easy surfaces until you are satisfied that s/he is steady.

- Continue to be observant when your person is maneuvering with the cane. You never know when s/he may get disoriented and fall.

- It is possible that learning a new technique or tool, even something as simple as a cane may just be too difficult for your person. Consult with your physical therapist for other solutions.

Needs to use a walker

Hopefully you will have had a consultation with a PT to fit the walker. Walkers come with or without wheels. The PT is responsible for training you and your person in the proper use of the walker.

- If you have not had a consultation with a physical therapist, ask your MD for one.

- Your person should practice with the walker on easy surfaces until you are satisfied that s/he is steady.

- Continue to be observant when your person is maneuvering with the walker. You never know when s/he may get disoriented and fall.

- As simple as it looks, a walker still requires coordination and practice.

- It is possible that learning a new technique or tool, even something as simple as a cane may just be too difficult for your person. Consult with you physical therapist for other solutions.

Needs help getting in or out of bed

As Alzheimer's disease/dementia (AD) progresses motor weakness and inability to understand the steps in performing a task are to be expected.

- Elevating your person's bed may be the easiest way to deal with this. If your person can just sit on the bed and swing his/her legs in or out of bed, it will be easier.

- Be there at bedtime to help your person.

- Give directions first and see if your person can perform the task alone.

- If simple directions do not work, then offer physical help.

Needs help getting in or out of chair

As AD progresses motor weakness and inability to understand the steps in performing a task are to be expected.

- There are assistive chairs and devices that fit in chairs that tilt to allow your person to get into and out of a chair at one height. Your person settles in for sitting, then tilts out to stand back up.

- Make sure the family knows that your person needs help getting up.

- Be careful of your own back when assisting your person out of a chair. There are books that can instruct you in transfers or consult a physical therapist (PT).

Notes:

• Special belts are available to use when assisting people with transfers. They can be obtained at a medical supply store. Ask for advice on how to use it.

Cannot walk spontaneously or independently

A person with AD may stop walking not because the muscles stop working but s/he may simply forget how to move his/her feet in the proper sequence.

• Give simple reminders, such as putting one foot out followed by the other. This may jump-start your person and s/he may walk without difficulty until the disorientation starts in again.

• Take your person's arm and walk with him/her. The forward motion may be enough cuing for him/her to mimic your steps.

• When prompts fail, it is likely your person is becoming late stage AD and is less responsive in general. Wheelchairs with transfers may be the next step.

Needs to use wheelchair

If your person needs to regularly use a wheelchair, it is best to get a fitting from a physical therapist. If the use is just intermittent (like at the airport or grocery store), then you will probably just make due with what is available.

• If AD has made your person non-ambulatory, you should keep the wheelchair simple. It will be difficult at that stage to learn to use the wheelchair.

• Some persons with AD merely propel themselves by moving their feet and "walk the chair" rather than move the wheels with the handrails.

• At some point, you will probably have to manage your person's maneuvering in the wheelchair.

Requires transfer from wheelchair to bed or chair

If you are not physically able to comfortably help your person make the transfer, there are assistive tools that will effectively hoist your person into bed. Again you should ask for a physical therapy evaluation to get good instruction in using these devices.

• The PT can come to your home for a few visits under Medicare to show you proper techniques.

• The PT can also instruct other helping family members or friends in the techniques as well.

Is bed bound

This is the most difficult level of care-requiring 24hr. tasks. Many people are not able to do this in their own home. Do not feel guilty if you have to draw the line. Your own health is very important. If you are going to provide full care in your home, you will need help from your network and instruction from Home Healthcare.

• Ask your MD for a referral to Home Healthcare.

• There are a number of books on the market that go into great detail about providing this kind of care.

Notes:

WANDERING

General

Wandering and/or pacing are very common during certain phases of Alzheimer's disease/dementia (AD). The behavior may go away then return during a later phase. Wandering is when your person tends to walk around aimlessly or seems directed toward a destination that s/he cannot explain. It is generally the result of some cognitive disorientation that happens when your person is unable to process sensory and cognitive information as s/he once did. Wandering can be an expression of stress; or your person may be searching for something-a vague sense of security, a deceased parent, for instance. Your person may be saying, "I feel lost;" and s/he is looking for the familiar. Your person may be asserting his/her will, trying to maintain some sense of autonomy. Or wandering could just be a response to boredom. Either way it can place your person in some danger if, s/he wanders outside and gets lost. This is a hard one for many families. You want your person to be independent, and you can underestimate the danger your person may be in when s/he is allowed to walk the city or countryside alone.

Tendency to wander

• Observe your person; get to know his/her patterns. You may be able to watch and s/he may be oblivious to you. You will be developing "that eye in the back of your head" sense that you had if you parented small children.

• Ask your person how s/he is feeling if s/he appears restless. Try distraction with activities (folding clothes, housework, shop activities or snacks to make him/her feel calmer).

• If your person is happy and just seems to be exercising or exploring, join him/her for a bit.

• Ask him/her where s/he is going. If it is a reasonable destination like the bathroom or his/her room, take them there. If it's not, redirect your person's attention.

• Take your person outside, to a shopping mall or somewhere to walk. The exercise will help him/her rest and the change of scenery will redirect his/her thinking.

• Don't wear them out. Fatigue can produce restlessness as readily as lack of exercise.

• Just sit down with your person and read out loud.

• Redirect the conversation to mealtime or ask if s/he is feeling better these days.

• There are geriatric equivalents of high chairs. They are high backed padded chairs with a waist high tray for doing activities. This will effectively immobilize your person without using restraints.

• Some persons with AD wander away while out in public. Get him/her a "Medic Alert:" or other ID bracelet or necklace that alerts people to his or her memory impairment and gives a contact number. If it's a bracelet make sure it's adjustable if your person tends to lose weight.

• Enroll in the Alzheimer's Association Safe Return program.

Notes:

• Keep a close eye on your person while out in public places. Just like a small child s/he can wander off while your back is turned. It can be easy for your person to get lost in a large department store or a mall.

• Write out an instruction card for your person's pocket. Tell him/her to refer to it if s/he is lost. It will tell your person to be calm, give a number to call and have his/her address in case someone else needs to direct him/her.

• Get a set of cheap walkie-talkies. Give one to your person when you go out. Teach him/her to use it or to respond when s/he hears it. You can even teach him/her to get someone else to respond to your call if s/he is confused.

• Some people won't go outside without their purse, wallet, and a favorite article of clothing such as a coat or hat or shoes. If this is the case, put these things out of sight until they are needed.

Wanders at night

Night wandering can be hard on a family. If you are not sleeping in the same room with your person, s/he can burst into your room and scare you half out of your wits. Or your person can get restless and wander out of your house into the night. As you might imagine, this is an extremely dangerous situation. The first time my mother-in-law did this she got a mile or so away and burst into someone else's home. Thank God they did not shoot her and their dogs did not attack her. The police figured out whom she might be related to and returned her within a few hours. She also tore off her Medic Alert bracelet so that was not helpful in this instance. Our family slept through the whole incident until the police showed up with sirens blaring. Up to this point, my mother-in-law had been very nervous about going out at night. After this episode the children called her Houdini Grandma because she had managed to undo the lock. Many of us have to learn the hard way!!

• Make sure your person gets some degree of activity during the day and does not take long naps that might hinder his/her ability to sleep at night.

• Leave a light on in the bathroom so s/he may be less disoriented if up at night.

• Make sure you lock the doors to the outside so s/he can't get them open. Even install safety locks that will be difficult for a memory-impaired person to learn (often dead bolts that are high out of the visual field).

• Take away his/her shoes at night and keep them in your room. This can discourage those with delicate feet from sneaking outside at night.

• It may sound cruel, but you can try locking your person in the bedroom if s/he has access to a bathroom or are using diapers. An easy way is to reverse a locking door handle on a bedroom door so it locks from the outside. If you do this be sure you have a fire safety plan in case of an emergency.

• If additional locks are impractical, consider a door alarm that will alert you if the front door or even your person's bedroom door opens.

• Install a motion detector light outside the house. It will alert you if your person "escapes" in the dark and, it may keep them from falling.

TRANSPORTATION & DRIVING

General

Driving is a very fragile skill in dementia. Your person will have lost quite a few fundamental skills before you observe the losses. Driving is very disorienting and even if your person has not gotten lost while walking alone s/he may get lost while driving. When persons with Alzheimer's disease/dementia (AD) have been able to describe the problem in early stages they say it is sudden and intermittent. You just look up and all of a sudden nothing seems familiar. You may not even be able to figure out how to get out of a parking lot. Persons without AD can experience a moment of such disorientation while preoccupied with other concerns. With persons with AD, the problem may persist for hours before things return to normal. All this makes driving for persons with AD so dangerous. Of course driving is often the one thing that a person with AD is unwilling to give up. This causes great concern for families and health professionals.

Can drive

If your person has been diagnosed with AD, it's time to start finding ways of limiting and supervising his/her driving. As the disease progresses, your person will have episodes of memory loss or perceptual aberrations. If this happens while driving, your person could become lost or have a serious accident. The problem will be intermittent in the beginning. If your person has been diagnosed in early stages, it will be good to have a conversation with your MD about driving. If possible your person needs to become aware of the problem and to alert you if and when the difficulties start.

- Do not encourage long drives alone.

- Have your person use the car for short trips.

- Ride with your person and observe his/her abilities first hand.

- Encourage your person to report any problems s/he is having. Know that this is difficult because s/he may fear having to quit driving.

- It may be possible that your person may not understand the problems s/he is having. Observation is always important if your person is inarticulate or suspiciously brief about describing problems.

Able to use a bus or taxi

If your person can still handle public transit, then this can be very helpful to you in the early stages. As with driving, there is a lot of sensory input coming into a brain that is losing its ability to process complex information. It will be hard to determine when the process will unravel, but it inevitably will.

- Provide your person with bus tokens, so that s/he does not have to make change under stressful circumstances

- Provide your person with a card that contains his/her address, phone number and instructions on what to do if s/he is lost.

- Begin an account with a local taxi company for your person.

- Be observant. Travel on public transportation with your person and verify his/her skills.

- As soon as you suspect that your person may be losing the ability to travel, stop his/her trips alone. You do not want to have to launch a manhunt for your person in a big city.

Notes:

• Obtain a "Medic Alert " bracelet and enroll in the Alzheimer's Associations "Safe Return" program.

Cannot safely drive but will not give up driving

This is a big dilemma in AD. So many people do not want to give up this major symbol of independence.

• Have any important people in your person's life (that s/he might listen to) talk to him/her and try to gently convince him/her that driving is no longer safe. Do not use hostile or confrontative means or your person will most likely become more obstinate.

• If your person will not voluntarily give up, then you must somehow disable the car.

• This can be difficult with someone who knows a fair amount about car repairs. Start with the distributor cap. Without it the car won't start. If you don't know how to do this, then ask a neighbor or friend.

• Some persons with AD will just call the car shop. Try to get to the car shop first and ask them to tell your person that the car cannot be repaired for now, or it may need a special part that has to be ordered.

• Hide the keys and say they are lost.

• If this is your primary vehicle, make up something about having to drive together to conserve gas!

• Use whatever tactic works. If the car belongs to your person and you can remove it from sight for "repairs", then do that.

• If the car can stay out of commission long enough, your person will begin to lose the ability to manage the car.

Must be driven places

Actually this can be a relief if you have had a difficult time with your person about giving up driving. At least you know your person is safe, not lost, or in peril of an accident. Some of the precautions here are similar to those you would use with a child (**retrogenesis** – reverse development).

• Always use a seatbelt on your person. You never know if your person might get disoriented and try to get out of the car.

• Depending on his/her level of AD you may want to have your person ride in the back seat to avoid any random, sudden (and possibly dangerous) activity.

• Never leave your person alone in a car. S/he could wander off, release a parking brake, or some other dangerous activity that no one has thought of yet.

EXERCISE

General

Exercise is so important for maintaining a healthy mind, body and spirit. There is now quite an emphasis on it in the popular culture. Unfortunately many people still do not incorporate even the minimum levels into their daily lives. No matter where you are starting from, any time is the best time to work on an exercise program. If you are a regular exerciser then you will want to work out a program so that you can maintain what you have been doing. You can work on programs that allow you to exercise

Notes:

together with your person with Alzheimer's disease/dementia (AD).

Likes to exercise on a routine basis

The better shape your person was in before s/he contracted the disease, the more likely s/he will engage in exercise as the disease progresses. Exercise will keep the joints supple, aid in circulation, and in burning off excess energy.

• Try to maintain any current exercise regime for as long as your person is able to.

• Consider any safety issues that may come up in the near future, such as being able to jog alone or the ability to work with exercise machines.

• If s/he works out in a gym, you can discreetly let the floor managers know that s/he may possibly experience some disorientation. They can report back to you if there are problems.

• Consider exercising with your person to keep an eye on his/her abilities. It could be good for the both of you.

• If the more complex exercises become too difficult for your person, try walking once a day with your person.

Seldom exercises

Exercise may be a key to keeping your person healthier, longer. People who have eaten a lot because of stress may find that the forgetfulness is a blessing in disguise, in that they forget to worry. With a healthy diet and exercise, your person may lose weight (if they're overweight).

• Now is not the time to start anything too elaborate. Your person may have a difficult time negotiating the gym if s/he never spent much time in one.

• Consider a simple walking program for the two of you.

• Take walks at the local zoo, park, or nature sanctuary. These walks can bring both of you some peaceful time. Try to leave your worries at home and relax with your person.

• As your person becomes more childlike s/he may re-experience the natural environment with new eyes and renewed pleasure.

• Encourage your person's interest in nature. It may bring a soothing spiritual connectedness with the earth and its rhythms.

• Bring your person to a gym to walk.

• Bring your person to an outside track to walk around.

Likes to play sports

Your person may be able to pursue sports and other recreational activities that have been life-long habits. Skill in tennis or golf for example may persist as long as they are physically able to perform the activity. AD losses tend to be very individual.

• Maintain some gentle contact with his/her favorite sport in a non-threatening way.

• Stay away from competitive matches where your person may end up feeling humiliated by an inconsistent performance.

• Discreetly let others in the game or match know what is happening, so that they can be sensitive to your person's needs and feelings.

• The idea is for your person to continue to have a good time at something s/he was good at.

Notes:

• If things start going down hill and your person expresses a lot of anger and frustration, phase the sport out.

Recreational activity of choice

Any activity, which your person enjoyed or was proficient at, is fair game for keeping active. Take the attitude that your person's work life is over and s/he should have as much fun and enjoyment as s/he can and you as the caregiver can tolerate. This is an area where friends can help. Perhaps an old friend who loved to have fun with your person can still take him/her out occasionally-especially in the early stages. This can give you respite. Activities that create their own context are especially good and they don't require any special rules to understand.

- Playing music
- Dancing
- Gardening
- Bicycle riding (Consider the safety issues.)
- Swimming
- Going to outdoor events that require moderate walking
- Fairs
- Concerts
- Parades

Has no interest in exercise or sports

Exercise is important so any small things that you can do to get your person moving are worthwhile.

- Walking is still your best bet. Your person can get interested in the surrounding nature.

- If your person can't do 20 minutes of exercise, do 5 minutes, then rest and play again.

Movements are severely restricted

There are a number of chair exercises that your person can do if s/he can follow directions.

- Lift weights from a chair.

- Buy a book on chair exercises and do them together or make a game out of it.

- Consult a physical therapist to see what kinds of activities or movements might be reasonable and helpful.

- If your person shows no interest at all in any recreational movement, sneak a walk in whenever you can.

- If it all gets to be too much of a hassle then settle for letting your person be a couch potato.

- A peaceful house is better for your mental health.

Notes:

MANEUVERING IN THE HOUSE

General

This is actually a huge area for consideration. While your person still has his/her faculties, it is important to establish routines and patterns of activities. It is likewise important to change as little of your person's surroundings as possible.

If it is your spouse with Alzheimer's disease/dementia (AD) and you intend to provide care in the home, don't make any major changes to the house or the interior environment (except safety changes.) The more familiar your person is with his/her environment, the easier things will be. If it's Mom or Dad with AD and you are moving them in with you, be prepared for their confusion to be compounded until s/he gets used to the new surroundings. The unfamiliar is the challenge.

Able to negotiate the house and recognize rooms

This is an important aspect of functionality. As time goes by this familiarity will become fundamental and supportive.

• Make as few changes to your house (except safety changes), especially if your person has been living there for a number of years.

• Make changes for safety, like railings etc. as needed. Explain these changes as best you can to your person. S/he may just adapt because the change (like the railing) is so easy to use and reflexively hold onto.

• If your person is in early stages, you can consult with him/her on changes that could be helpful and make your home safer.

Unable to recognize or find own room

This is a middle stage issue and needs to be addressed in a supportive manner.

• Put a sign on the door with your person's name.

• Put a sign on the door with the a generic label like "bedroom."

• Put a picture of the function (a bed for the bedroom) on the door.

Unable to recognize other rooms in house

It's important that your person at least be able to find the bathroom.

• Put a sign on the door.

• Put the name of the room in printing on the door.

• Put a picture of the function (a toilet for the bathroom) on a door.

• Put a picture of your person whose room it is on the door. Be aware that at later middle stage, your person may begin to lose contact with what s/he looks like. Due to **retrogenesis** – reverse development, s/he may picture him/herself as a much younger person. This is what causes persons with AD to become angry at mirrors. They no longer recognize the older image that is reflecting back at them.

Wanders into other family members rooms

This is a disconcerting issue for "sandwiched" families. If your person wanders the house and bursts into your room at night, it can be incredibly disturbing.

• If privacy is the issue, keep your door closed and locked. Your person may still bang on your door or make a disturbance.

• Refer to the section, **Wanders at night** (page 76).

Notes:

Cannot recognize outside of house

This is typical of the wandering period. When a person with AD loses track of where s/he is, landmarks also fade in and out. Vague recognition of familiar places may linger, but names and contexts will be lost.

• If your person does not recognize the look of your house or the recognition is intermittent, this is a fairly good indication that s/he can get lost if allowed to walk in the community alone. Now would be the time to have your person walk only with a companion. See section on WANDERING (page 75).

Imagines there are other rooms or another floor in the house

This is not uncommon. Persons with AD often become confused about where they are. Your person may believe that s/he is in a house previous to the one in which s/he is currently living. The memories that come alive in your person's brain can take him/her back to childhood homes. S/he may be looking for rooms from another time.

Don't argue; respect your person's reality even if it is different from yours.

• Go with the flow, talk to your person about the rooms or houses s/he may be seeking. Ask who else may live there. Pieces of your family's history could be revealed.

• Allow your person to search as long as s/he does not become agitated or emotionally out of control.

• If your person is not frantic about finding the other rooms, walk with him/her and listen to the story.

• Be creative and see what works. Walk your person through a few doorways and suggest you both have arrived on the next floor. This may satisfy your person. You will be surprised what works in the "Alice in Wonderland " world of dementia reality.

• If your person becomes agitated, as always distract. Offer food or another activity that will take his/her mind off of the search.

• Offer to look later for the room and any items that seem important to your person.

• Keep your voice calm even melodic when guiding your person to another activity.

SAFETY

General

Safety must be foremost in your mind, as your person becomes more childlike (**retrogenesis** – reverse development). In many ways you have to build your safety plan in reverse of childhood, getting more complete as your person's disease progresses. It is all too easy to overestimate the skills of your person because s/he looks more in command than s/he is. This is especially true when it concerns driving and walking alone.

Safety is not yet an issue

This is the time to begin to devise your plan. When Alzheimer's disease/ dementia (AD) takes hold, it is a bit like having a tall, strong, clever 2-year-old in the house. You will eventually have to secure all poisons and cleaning fluids.

Notes:

• Adapt the house but don't make wholesale changes that could confuse your person. This could depend on the level of clutter in your house. Even if it may be slightly more confusing, lowering clutter is still a good idea.

• Simplify the home; but keep familiar possessions for comfort.

• You can remove things like throw rugs that can soon become a hazard encouraging falls.

• Allow your person to behave as normally as possible, but be on the out for potential problems.

• This is a good time to remove firearms from your house or store them securely, unloaded, under lock and key. Store the bullets in a separate locked place, and make sure you have the only keys. People tell stories of men pulling guns on their families. As you can imagine this can a very dangerous situation. Be especially careful with involving police. They have been known to shoot and kill demented elderly people who are waving guns and threatening them.

Beginning to use household appliances unsafely

It is best to allow things to proceed as normally as possible. If your person is beginning to have problems using appliances don't completely stop the use, start with supports.

• Observe the task.

• Define the difficulty.

• Try offering support for whatever parts of the task that your person cannot perform by him/herself.

• Offer the support for as long as it works

• Over time more support will be needed; add it slowly as necessary

• When your person can barely function with the appliance, or when it becomes to much trouble to offer support, or if the use of that appliance becomes a safety issue, phase out his/her use of the item.

• If your person becomes upset about not being able to use a certain appliance, make up some excuse perhaps that it's not working. The purpose is to let your person down gently.

• Never demean your person about his/her diminishing skills

• Unplug appliances until ready for use to keep your person safe from unsupervised use.

The house needs to be safety/dementia proofed

As stated before, you will eventually have to safety proof your house. Your person may at some point lose the ability to tell the difference between the edible and the poisonous. Just like a small child, s/he will need to be protected from making a mistake that could harm him/her. Other dangers come from the potential for slipping and falling.

• Remove or secure poisons: household cleaners, bleaches, ammonia, and solvents.

• Remove or secure pills and medicines.

• Install grab bars in the bathrooms.

• Put away all the sharp objects, especially knives.

• If you have not already done so, this is the absolute time to remove weapons from your house or make sure they are securely locked up

Notes:

and you have the only set of keys. Disarm them as well, removing bullets etc. If you ask, people will tell you stories of men especially pulling guns on their families. As you can imagine this can be such a dangerous situation. You also have to be careful with the police as well. They have been known to shoot and kill demented elderly people who are waving guns and threatening them.

COOKING

General

Cooking and preparing a meal can be a lone activity or a social one. Persons with Alzheimer's disease/dementia (AD) need involvement in family activity to feel useful, helpful, and loved. Cooking for persons with AD is no longer a learning experience, but one that can stimulate brain function and preserve skills. Whether or not s/he has cooked in his/her life, there may be warm memories of family life associated with the kitchen and sharing meals. Almost everyone has done some kitchen-related chores. Cooking with your person while s/he still has the skills is great for interaction and to simply keep an eye on him/her. Cooking tasks should be supervised because for someone who does not have all of his faculties, the kitchen can be a dangerous place. There are knives and other sharp objects. There is electricity and water in proximity to each other. There are foods and poisons (as in some cleaning products). And probably most dangerous is fire from stoves, ovens and other electrical appliances. On the other hand, the kitchen is a wonderland of simple, repetitive activities. There are familiar sights and smells that help recall memories. Any activity done together is a social activity for your person.

NOTE: Elderly men have not been as culturally adjusted to cooking as women. This is changing. If you can get him involved in cooking it can be good, if not, refer to the **WORKING AROUND THE HOUSE OR SHOP** section (page 97) of the book for other ideas for keeping your person busy.

Can cook without help

Cooking without help is both a blessing and a curse. For a person diagnosed with AD, it is only a matter of time before his/her attention wanders.

• If your person enjoys the kitchen, hang around and watch how well s/he cooks.

• Be alert for lapses in safe behavior or distracted behavior.

• Try to be unobtrusive as you work along side and make things safe.

• Offer to help them cook so that you can be useful and supervisory at the same time.

Can cook with help

If your person cannot safely cook by him/herself, volunteer to be a helper. Or if it is your kitchen, suggest your person be the helper and assign them various simple tasks.

• Start as a helper to your person and transition to cook and guide. Make this transition as painless as possible, while still keeping safety in mind.

• Compliment your person on his/her skills as you offer increasing levels of assistance.

• Emphasize the idea of cooking as a team.

• Observe his/her knife techniques before turning your back on that activity, or you can chop vegetables next to your person and observe his/her knife skills.

• Observe his/her abilities with any appliance as they all have dangerous aspects.

• Work with your person to perform the activity rather than going on to another task.

• Give yourself more time to cook so that you can work with your person as a team without pressure.

• Encourage your person to locate kitchen items like pots and pans. This will help his/her organizational skills while they still exist. If it takes some time, just let him/her do it at his/her pace. Step in if s/he expresses frustration.

• Emphasize the positive in what your person is able to do, not the negative. Do not blame or chide.

• Expect things to get broken, so switch out glass for plastic or metal.

• Try not to give your person access to heirloom dishes and the like. If s/he heads that way distract him/her and give the task of handling the dish to someone else.

• Guide your person toward less dangerous tasks (measuring for example).

Can turn on and off stove and oven

This is a delicate and dangerous area of function. Many persons with Alzheimer's disease/dementia (AD) nearly start fires because they forget they even started cooking until (hopefully) they smell smoke and turn the stove off.

If your person uses the stove properly, take some care to watch behind him/her to make sure the stove really is turned off. While there is generally less danger from the oven, it can be a source of severe burns if your person doesn't use potholders or oven mitts.

• Observe this skill routinely and if possible just get into the practice of being in the kitchen when your person is trying to cook something. Always let him/her do as much independently as possible.

• Offer gentle reminders if your person looks like s/he is heading for trouble.

• Also keep an eye out to notice if s/he handles hot pots properly.

• Make sure potholders and oven mitts are readily accessible.

• Make sure the pilot is on and ignited in a gas oven when your person turns up the heat. Walk past the stove, open the door, and double check. Pretend you're admiring his/her work.

Can turn stove or oven on but forgets to turn off

Logically your person should never use the stove without supervision.

• An easy fix is to remove the knobs, and though inconvenient, it will afford a great margin of safety.

• If you are able, have a gas cut-off valve installed and keep it off when you're not around.

Notes:

• If the stove is electric, have a master switch installed and turn the stove and oven off.

• Put decorative covers over the burners – your person might not notice it's a stove any more.

• It may also be time to consider putting up some kind of barrier at the kitchen door. This may be something to try, but s/he may still get past it if there is something s/he desires.

• Disable or put away all appliances that may be dangerous if turned on unsupervised.

Can pour boiling water

This is another potentially dangerous activity that will require supervision.

• Observe your person's abilities to perform this skill. Look for potential dangers.

• Make certain potholders and/or oven mitts are available.

• Begin the practice of being in the kitchen when s/he wants to work with boiling water.

• Find other ways to get what s/he wants without boiling water. Coffee is most often made by the pot so, you can get a small coffeemaker. Tea can also be fixed in these small appliances. This will mean the hot beverage only has to be poured once.

• Offer to make the beverages for them as a courtesy.

• Phase out this as a lone activity as soon as you feel comfortable doing so.

Cannot pour boiling water

• Remove any teakettles.

• Disable the stove as stated in the section above.

• Remove any coffee or tea makers from sight until your person no longer seems to remember how to use it.

• Fix all hot beverages for your person being very careful to let them cool some before serving, in case the drink is spilled.

Can read recipe and measure ingredients

Again you want to encourage independence and practice with skills, but keep a watchful eye if you want things to turn out okay. These are familiar, comforting and rewarding activities that make your person feel better about him/herself.

• Watch this closely for indications of slippage. These are complex sets of activities requiring reading, math, memory and motor skills.

• Choose recipes that have a margin for error.

• Just let your person measure items for the fun of it. You do not have to use what s/he measures.

Cannot read recipe and measure ingredients

This need not be a big stumbling block. Take cues from your person. If s/he is not concerned, then go on with things; but if this inability is upsetting, then make some accommodations.

• Ask your person to keep you company as you cook.

• Give him/her small tasks.

Notes:

• Offer to make your person's favorite recipe and give directions that make him/her feel involved as if s/he was still able to read the recipe.

• Stirring is a good activity for persons with dementia and children.

Can use microwave

Watch this. Keep activities on the microwave simple, and observe how it is used and if there is any confusion. It's easy to overlook this appliance, but, if your person puts the wrong thing in the microwave, it can be dangerous, costly or messy (or all three). Make sure s/he knows which buttons to hit so s/he doesn't cook the food too long. Microwaved food can get dangerously hot.

• Make sure your person removes the foil from anything s/he sticks in the microwave; this can blow out the magnetron in the machine.

• Make sure s/he does not put in whole eggs – very messy.

• Make sure your person uses the right type of dish or plate. Glass dishes can get very hot (paper plates are best).

• Begin to supervise this activity even before your person loses his/her ability.

Cannot use microwave

As with any appliance that represents a danger to your person disable it and hope s/he simply forgets to want to use it.

• Unplug the microwave after each use.

• Hide the cord behind the machine.

• Don't let your person see you plugging it in.

• Declare the machine is not working when it doesn't come on when your person tries to use it. Don't offer to trouble shoot just offer them an alternative way to heat or get food.

• Perform the task for your person and unplug the machine after each use. Tell him/her that the microwave works intermittently.

Able to set the table

This is an excellent activity for your person. It will keep him/her involved while you attend to something else.

• Think about retiring your good silver, plates and glasses, even if your person is very careful. You don't want to tempt fate. This will bring down your stress level considerably.

• Use only unbreakable items for your table settings.

• Don't get hung up if anything is in the wrong place. If you must, discreetly change it and go on.

• Compliment your person on his/her work.

Cannot manage table-setting at all

This is not hard to work around. Allow your person to do whatever small parts of the tasks s/he might still be able to do.

• Let your person hold and hand you the various items for the table.

• Let your person work with one of your children to set the table. Make sure the child understands to make it a game and not to get angry if Grandma or Grandpa makes a mistake or drops something.

• Let your person sit at the table and talk to you as you set the table.

• Have your person fold napkins or the like if s/he still can.

Notes:

• Give them a piece of bread or small snack to eat as you work.

Can use small electrical appliances

• Most electrical appliances in the kitchen require multi-step operation. Toaster ovens require setting temperatures and usage (toaster and oven). Blenders require knowing the appropriate speed and what the speed buttons are. Food processors have sharp blades and must be closed tightly and sometimes require twisting and feeding motions simultaneously. Take these things into consideration as you allow your person to operate these appliances.

• Observe this skill routinely and if possible just get into the practice of being in the kitchen when your person is trying to cook something. Always let him/her do as much independently as safely and reasonably possible.

• Offer gentle reminders if your person looks like s/he is heading for trouble.

Cannot use small electrical appliances

Encourage your person to hand mix things or do it for them.

• Unplug the appliance after each use.

• Hide the cord behind the machine.

• Don't let your person see you plugging it in.

• Declare the machine is not working when it doesn't come on when your person tries to use it. Don't offer to trouble shoot; just offer them an alternative way.

• Remove small appliances from sight so they are out of mind.

Can slice and chop with a knife

• Observe his/her knife techniques before turning your back on that activity or you can chop vegetables next to your person and observe his/her knife skills.

• Make sure they are using a chopping board and that there is not clutter around where they are working.

• If your person has a tendency to cut him/herself, think about curtailing its use.

Must be supervised when slicing or chopping

• Stand by as your person works and observe that s/he is using the utensil properly.

• Think about getting a manual-chopping tool, which shields hands from blades.

Cannot use a knife

• Lock up the knives or put them out of reach.

• Distract your person if s/he wants to cut or chop vegetables etc. Tell him/her you are going to cut or chop using an appliance. Let him/her watch or assign another chore.

Cannot perform any cooking tasks

Hopefully by this time your person doesn't care about the kitchen anymore. If s/he still seems interested, find distractions that are safe to do in the kitchen while you work.

Notes:

• Let your person sit at the table and eat a snack or work on some drawing or reading activity as you work.

• Try to make the kitchen as safe as possible so that there are no dangers for your person when you are not right there to watch him/her.

• Discourage your person from going into the kitchen when you are not going to be there. Keep him/her occupied in the area of the house that you are going to be in (this is not too different from what you need to do when you have toddlers in your house. Call on those skills if you have had children).

• Once s/he has forgotten how to do any of the tasks, the kitchen will probably be safe for him/her.

WASHING DISHES

General

For people who have kept house for much of their lives, dishwashing carries a lot of household and rote skill memory. It is about nurturing, sustaining and keeping a family clean. Any activity done together is a social activity for your person.

Can wash dishes without help

Dishwashing offers a great opportunity for your person to be both useful and occupied. Playing in the water and soapsuds is tactilely satisfying. Another part of the task is being able to know the difference between clean and dirty dishes. Determining the difference between clean and dirty dishes can require that subtle distinctions be made. Your person can take pride in doing a chore that is necessary and important to the household.

• Observe this skill routinely, and if possible just get into the practice of being in the kitchen when your person is washing dishes. Always let him/her do as much independently as possible.

• Offer gentle reminders if your person looks like s/he is heading for trouble.

• Also keep an eye out to notice if they handle hot water properly.

• Turn down the temperature on the kitchen water heater if it comes out of the tap excessively hot.

Can wash and dry dishes with supervision

The emphasis continues to be on preserving the skills. At this point some of the parts of the task have come unraveled and need oversight to remain intact. One aspect is the inability to distinguish between clean and dirty dishes. This indicates that your person has lost some of his/her powers of discrimination. His/her perceptions of the meaning of clean and dirty as well as the ability to apply this appropriately has become impaired. You will now need to keep an eye on the situation. Doing dishes with another person is often a pleasant way to get a necessary job done. Your person will have the sense that he/she is helping and being useful. Having your person dry and stack the dishes helps them keep some level of organizational function. You can use this dishwashing time to inventory your person's state of mind in a relaxed setting.

• Always let him/her do as much independently as possible.

• Offer gentle reminders if your person looks like s/he is heading for trouble.

Notes:

Behavior's List: Physical – Washing Dishes

- Set the water handle to the place where your person can get warm water for filling the sink or rinsing.
- Fill the sink for your person and add the soap.
- You can trade tasks- -dry while your person washes and vice versa.
- Switch to plastic dishes and glasses if you think your person with dementia may drop and break regular dishes.
- Offer gentle reminders if your person is skipping a dish or trying to dry or rinse an unwashed dish. Also watch to see if s/he tries to put away dirty dishes.
- Do the rinsing your self, or make the project a team effort where your person washes and you rinse.
- Watch the use of the sink drainer to keep your person from rinsing food straight into your pipes.

Puts away dishes in wrong places
This can be frustrating especially when you start to look for misplaced things. Again this calls for supervision.

Rearranges kitchen drawers or cabinets
Your person is likely doing this when you are not in the kitchen. Perhaps it is time to supervise his/her activities in the kitchen.

- Go to a thrift store and buy some old flatware and a silverware drawer tray. Set out the box of unsorted flatware and let your person sort to his/her heart's content. If your person's memory is poor, s/he can do this every day, with no impact on the household.

Can tell the difference between leftover foods and trash
Let your person do the task of putting away leftovers. It will keep him/her engaged and make him/her feel useful.

Cannot tell the difference between leftover foods and trash
- Break the task up into the parts your person can and can't do.
- Have him/her bring you all the food.
- Tell your person that you are in charge of deciding what is trash and will tell him/her which things to throw away.
- Let your person scrape food off of the used dinner plates into the trash. Just be observant that the dishes or utensils don't suddenly end up in the trash.

Tends to toss good food or utensils in the trash
- Don't let your person use the trash can alone.
- Have him/her bring everything to you.
- You throw things out.
- Look in the trashcan every now and then to see that they are complying.
- It may be time to supervise your person's time in the kitchen.

Cannot perform any washing tasks
You can still have your person remain in the environment while you are washing dishes and perform any small tasks s/he is still capable of.

- Allow your person to sit at the kitchen or dining room table whichever is closest to you.

Behavior's List: Physical – Cleaning

Notes:

• Give him/her a small snack to eat or an activity to do. It can be a household chore like sorting silverware or folding napkins. It can also be an art activity like drawing or crayoning in a drawing book. The more child-oriented activities can be matter of factly offered and don't have to demean your person.

• Beyond mealtimes put the kitchen off limits unless you are supervising.

CLEANING

General

Cleaning is both a functional and symbolic activity. For a person with Alzheimer's disease/dementia (AD), the ability to organize their environment does good things for their self-image and sense of self-esteem. Bear in mind that your person may confuse the proper use of various cleaning products. You will need to be aware of any cleaning activities that involve cleaning products or sharp objects. Cleaning activities can be highly ingrained, have a fairly solid rote memory, and in some form or other last for a while. Any activity done together is a social activity for your person.

NOTE: Men have not been as culturally adjusted to cleaning as women. If you can get your person involved, it can be good. If not, refer to the **WORKING AROUND THE HOUSE OR SHOP** (page 97) section of the book for other ideas for keeping your person busy.

Cleans without difficulty

This is a skills preservation activity. Allow your person to participate whenever possible, especially if s/he has a desire to do so. You can still interest and occupy him/her in cleaning activities so that s/he doesn't wander or get into other mischief.

• Encourage your person to clean his/her room and the space over which s/he has control.

• Observe his/her skills at cleaning so that you can anticipate problems.

• Watch the use of cleaning products. Even though your person is high functioning you never know when s/he might make a mistake and spray furniture polish all over your mirrors or spray window cleaner all over your furniture.

Can clean with supervision

Your person has begun to mix things up a bit and needs to be supervised. This is not too much different because you have probably begun watching how s/he does things anyway, and you will need to continue to do so. Part of the trick is to find cleaning activities that are fairly harmless and will keep your person happily occupied.

• You can set up the vacuum for your person. It doesn't matter if s/he goes repeatedly over the same area. If s/he is occupied and feeling useful, that is what is important.

• Let your person push around one of those mechanical sweepers. They are harmless and can provide quite a bit of busywork.

• Let your person use a broom and dustpan. Somewhere this task will probably come unraveled, but it won't be a problem if you are prepared to sweep up after your person.

Alzheimer's Workbook   **91**

Notes:

• Let your person use one of the newer dusters that tends to pick up dust as it goes. These are usually made of plastic and won't be a problem if they accidentally get broken.

• Be watchful where your person is dusting. Avoid having him/her work where there are a lot of knick-knacks that have to be moved to dust. You never know when your person might drop something valuable. You may be able to trust him/her more if you are working right beside him/her.

• You could spray some polish on a table and let your person rub the polish in. Be sure to keep the can under your control or you could end up with polish in a number of unwanted places.

• You could do the same thing for windows.

• You could sprinkle cleanser on the counter top or in the sink and let your person do some scrubbing.

• Let your person use one of those newer plastic push mops. They are fairly harmless unless your person decides to take it apart. It does have a container of cleaning fluid attached. You can stay within sight of your person and s/he will probably not have the time to do much tinkering. The handle has a trigger for spraying the cleaning fluid. Your person can be shown how to use this. Not enough fluid comes out at a time to create a problem.

• By now most every activity should be supervised. Just work in the same room with your person to see that s/he has not gone off on some unsafe tangent.

• Praise your person for any help s/he has been able to offer.

Cleans excessively

Excessive cleaning is not normal and is likely to lead to bad consequences, especially if cleaning products are being used. Balance in all things is desirable for your person.

• Distract him/her with other cleaning chores or exercise.

• Take your person out to another activity and let him/her know s/he can finish later.

Can use vacuum cleaner

A vacuum cleaner provides good and useful repetitive behavior for the person with dementia. S/he can do little harm unless s/he runs into things. Watch to see that your person is careful. Praise him/her for his/her help.

Can use vacuum cleaner with supervision

• Make sure you plug and unplug the vacuum cleaner for your person.

• Watch the cord so it doesn't get tangled.

• Pick up any hard objects like keys or coins.

Cannot use vacuum cleaner

• Encourage your person to pick up larger things on the floor as you vacuum.

• Have your person use the mechanical sweeper and follow behind you.

• Give your person a smaller vacuum with the cord cut off and again have him/her mimic your behavior. S/he will most likely not notice

Notes:

that his/her vacuum doesn't make a noise. If s/he does get upset go back to the mechanical sweeper idea.

• If your person is fixated on the using the vacuum cleaner, s/he can be distracted to do other things or other chores.

Can sweep and use dustpan

This will be of help to you in the kitchen and other areas with hard floor surfaces.

• Keep an eye out for what is being swept up. Your person may not know valuable objects from trash.

Cannot sweep and use dustpan

If your person can hold the dustpan, have him/her hold it for you.

If not, ask him/her if s/he sees anything else or anywhere else that needs sweeping.

• Let your person mimic the activity. It really does not matter if s/he does it "right." This can apply to any activity. Just make sure s/he is safe and let him/her go for it.

Can wipe counters and table

• Provide your person with a pre-moistened sponge or cloth.

• Set the rules for cleaning products.

• Watch closely for inappropriate use.

• Just because your person uses the cleanser properly today, doesn't mean s/he will tomorrow.

• Put food away before your person works.

• Let your person wipe with a simple wet cloth without any soap or chemicals.

Can wipe counters and table with supervision

• You can sprinkle the cleanser or spray the cleaner and give your person the task of scrubbing or wiping.

• Ask your person to dry behind you. Give him/her a dry towel.

• Invite him/her to sit down and talk to you as you work or fix them a cold drink to sip as they watch you. This is especially good if your person is making a nuisance of him/herself.

• Let your person wipe with a simple wet cloth without any soap or chemicals. It does not really matter if s/he actually does the task "right."

Can empty trash

• Make sure the trash is not too heavy for your person.

• Make sure there are not glass shards sticking out or in the trash.

• Keep an eye on your person if s/he is going outside un-attended even for such a short time.

• Watch that s/he actually puts the trash in the right place.

• Supervise the activity, but still allow your person his/her independence

Cannot perform any cleaning tasks

• Have your person watch you clean.

• Talk to your person as you go about your chores.

Notes:

• Have your person sit at a nearby table and get him/her started on an art or play activity while you clean.

• Give your person some other task like folding laundry or sorting forks and spoons.

• Also set up an armchair table so that your person can keep occupied in an armchair as well.

WASHING CLOTHES

General

This is another household activity that can occupy your person and help preserve his/her skills, but it has sequential aspects and therefore is vulnerable to unraveling. If your person seems interested then this may be something you can involve him/her in. If your person has no interest then you may be able to have him/her act as a general helper, performing a few parts of the overall task.

Any activity done together is a social activity for your person.

NOTE: Men have not been as culturally adjusted to washing and other domestic chores as women. If you can get your person involved it can be good. If not, go to the **WORKING AROUND THE HOUSE OR SHOP** (page 97) area of the book for other ideas for keeping your person busy.

Can use washer and dryer

Let your person do his/her laundry and any household laundry that is not difficult to wash. It will keep him/her active.

• Sorting clothes for laundry is good brain activity.

• Observe his/her washing and drying skills so that you can anticipate problems.

• Observe his/her use of washing products so that you can anticipate problems.

Can use washer and dryer with supervision

There are some things to watch out for with laundry. The laundry room has a number of toxic and dangerous substances. As your person loses capacity, s/he will tend to make the same mistakes beginners make such as adding bleeding colors to whites or washing woolens in hot water, etc.

• Supervise the use of bleach with whites.

• Make certain your person is using the correct soap.

• Make certain your person doesn't put the wrong kinds of clothes in the dryer.

• Supervise the process and suggest working as a team so that you can cue your person to complete the overall task properly.

Cannot use washer and dryer

Once your person can no longer use the washer or dryer, you will need to place your person with Alzheimer's disease/dementia (AD) in a helper role.

• Encourage him/her to help carry the basket.

• Let your person remove the clothes from the dryer and place them in a basket.

• Work as a team giving parts of the task to your person.

Notes:

• Choose the parts that your person still does well and let him/her do those jobs.

• Praise his/her work and offer of help.

Can do hand washing

Doing laundry by hand can be a soothing activity. Playing in water and getting the soap on ones hands can be very satisfying. Again, caution must be exercised with the use of soaps and bleach. This activity is probably only useful if your person enjoys it. If not then there is no reason to go through all the bother; stick with other cleaning activities that your person enjoys.

• Set up a tub in your laundry room or kitchen with soap and allow your person to hand scrub some clothes.

• You will have to set up another tub for rinsing or take them to the sink.

• Supervise this activity because it could get messy.

Can do hand washing with supervision

If your person is used to doing for him/herself s/he may find reassurance in doing chores like hand washing his/her clothing. It helps with sequencing and supervised time is good for talking over the events of the day.

• Set up the tubs and do the wash together.

• Let your person perform any parts that can be done safely.

• This can be a social time just like other cleaning activities.

Can fold laundry

This can be such an easy and soothing activity for a person with dementia. At first s/he may be able to do it appropriately; but after while it can just be a repetitive activity using a basket of old clothes or rags kept around for just that purpose.

• The feel and smell of warm laundry can be comforting to your person.

• Familiar tasks can be a source of comfort for your person.

• Let your person fold it and put it away if s/he can;. otherwise you put it away.

• Be prepared for some unusual folds. Don't scold or try to instruct him/her on the proper way.

Can put laundry away

Allow your person as much latitude as possible; but remember, if there are children or a large family in the house, there are many places for laundry to go. Keep an eye on how s/he is doing with this activity. You many end up with your 10-year-old's underwear.

• Keep your person's jobs simple and easy for both your sake.

Can fold laundry with supervision

• Do the activity with your person. Be patient.

• If s/he piles or folds things incorrectly, just re-pile or refold them and put them away.

• Like with so many other activities, this is just something to keep your person occupied.

• Don't scold or try to instruct him/her on the proper way to fold.

• Allow your person to fold and refold the same laundry if it makes him/her happy, pleasantly occupied, and feel useful.

Notes:

Can put laundry away with supervision

• Have your person carry the basket for you or open the drawers.

• Make them feel useful but don't rely on them too heavily.

• Allow your person to be your helper and to do any part of a task that s/he can do easily.

Cannot fold laundry properly

• Give your person a shirt to pretend to fold as you fold the laundry.

• Talk about when you both were children and used to fold laundry.

• Give your person laundry to fold in anyway, and accept however it comes out.

• Let him/her fold and refold the same bit of laundry or keep a basket of old clothes and/or rags that your person can fold whenever an activity need arises. The whole point is to make your person feel happy and useful.

Can sort socks

Sock sorting is a great activity for pattern recognition.

• Save the socks for your person to sort.

• Even if s/he can do it alone, have him/her sort the socks first, you inspect, and then have him/her put them away. Don't be critical if the matches are not quite perfect.

• If s/he has mismatched a bunch, thank your person for his/her work and move on to the next task. You can correct the problem later, out of his/her sight.

Can sort socks with supervision

Sorting socks, even if done poorly can be good exercise for the brain.

• Save the socks for your person to sort.

• Work together looking for matches. You can make it into a game.

• If you have children around you can assign them to the sorting activity with Grandma or Grandpa. You can inspect before you put them away. Don't be critical if the matches are not quite perfect. You can correct the problem later, out of his/her sight.

Cannot sort socks

• Give your person one pair of socks to fold while you sort them.

• You can let him/her play with the socks as you sort them. Get your person to find certain colors as a game.

Cannot perform any laundry tasks

• Allow your person to perform an activity or have a snack near where you are working so that you can keep an eye on him/her.

WORKING AROUND THE HOUSE OR SHOP

General

In this day and age, men and women interchange jobs/ activities much more easily than in the past. The stereotypes of what is exclusively women's and men's work are melting away. For our current elderly population this can be more variable. In general, elderly women are more adapted to performing household chores, and elderly men are more comfortable doing household repairs and possibly working outdoors. Men may also have more competency issues about being able to do things "right" and may refuse to participate in activities. I have had requests from female caregivers to put in a section for more male oriented activities because their husbands/fathers absolutely refused to help around the house. Persons with Alzheimer's disease/dementia (AD) need to keep active to preserve function and to create some degree of happiness. This does not mean that your day has to be a mad rush, but a few "useful" activities can make your person less restless and easier to care for. A bored person with AD who just sits in a chair all day can be a problem. If this happens, first be sure to have your person evaluated for depression. Then try ideas that you think will suit your person. The ideas listed below as with the other sections in the workbook can be used for men or women. It just depends on what makes each individual happy.

Can work in shop/garage or make household repairs without help

Being able to perform any shop or repair task without help is both a blessing and a curse. For a person diagnosed with AD, it is only a matter of time before his/her attention wanders. There are many dangerous tools in a shop/ garage that your person may want to continue using. Evaluate carefully, and consistently assess what your person can actually safely do. It would seem to be wise to begin removing tools that may soon pose a danger. You would hate to find out your person can no longer use a tool because s/he had a disabling injury.

- If your person enjoys working in the shop/garage or making household repairs and still wants to do these activities, hang around and watch how well s/he works (or have a person with these skills do the observation).

- Be alert for lapses in safe behavior or distracted behavior.

- Try to be unobtrusive as you lend a hand and make things safe.

- You can offer to be a helper on whatever project s/he might want to work on. And then you can be useful and supervisory at the same time.

- Over time you can formulate a plan to actually make the shop/garage a safe work or "play" area for your person. Slowly begin removing dangerous items and provide easy access to safe ones. Provide materials for projects that only need hammer and nails verses a skill saw.

Can work in shop/garage or make household repairs but needs supervision

If your person cannot safely work in this area by him/herself, volunteer to be a helper. Or if it is your shop/garage, suggest your person be the helper and assign them various simple tasks.

- Start as a helper and transition to the project guide. Make this transition as painless as possible, while still keeping safety in mind.

Behavior's List: Physical – Working Around the House or Shop

Notes:

• Compliment your person on his/her skills as you offer increasing levels of assistance.

• Emphasize the idea of doing projects as a team.

• Observe his/her tool techniques before turning your back on that activity, or you can work next to your person so that you can see what is happening.

• Observe his/her abilities with any tools as they all have dangerous aspects.

• Work with your person to perform the activity rather than go on to another task.

• Give yourself more time to work on any project, so that you can work with your person as a team without pressure.

• Encourage your person to locate tools or items like certain size nails or bolts. This will help his/her organizational skills while they still exist. If it takes some time, just let him/her do it at his/her pace. Step in if s/he expresses frustration.

• Emphasize the positive in what your person is able to do, not the negative. Do not blame or chide.

• Expect things to get broken, so make sure s/he is not working on something you cannot afford to lose.

• Set up a parallel project to the one you are working on, with cheaper materials.

• Guide your person toward less dangerous work or tasks such as sorting, stacking or prep work.

• Now is the time to truly implement a plan to actually make the shop/garage a safe work or "play" area for your person. Slowly begin removing dangerous items and provide easy access to safe ones. Provide materials for projects that only need hammer and nails verses a skill saw. If this is your working shop/garage, perhaps create a safe "special area" for your person to work. This way you can keep all the more dangerous tools in your work area.

• At this stage, do not let your person work in the shop/garage without supervision unless you have created a completely safe area where no accidents can happen.

• If working in a shop/garage is not your area of expertise or pleasure, but it is for your person, then think about finding some family members or friends who can help work with your person a few times per week and implement some of the above ideas. It could be a team contribution to clean up the shop/garage and make it a safe area for your person. What a great gift that could be.

Cannot perform any shop/garage or household repair tasks

By this time your person may not care about the shop/garage anymore. If s/he still seems interested, find simple tasks that can be done in the house and will work as distractions while you are working at household chores.

• Let your person sit at the table and sort a special box of nails, screws or bolts. This can be made into a repetitive project by dumping sorted items back into the unsorted box to be sorted again—just like laundry.

• Continue to allow your person to work in the shop/garage as long as

98   *Alzheimer's Workbook*

s/he is still interested. If you do not already have a safe "special area" you can create one.

• If creating a big section "special area" is too complicated consider a small workbench. Your person can sit there and hammer nails into wood or sort nails, screws and bolts.

• Create your own simple projects for your person to do. You can visit a hobby shop to get inspiration and materials.

• Unless your shop/garage has been converted into a completely safe "play" area for your person, you will need to lock the door, so that your person cannot go in unless you are there to supervise.

• If your person is still interested but has no safe skills, you may want to consider toys such as a pounding bench (yes, the one with a hammer and pegs.) Only you will know if these will suit your person and not be demeaning.

• If regular toys won't work, you can go on-line and search for play boards at Alzheimer's/dementia sites. These items have several activates like locks or bolts that slide open. This is a nice table activity to provide some enjoyment for your person while you are working at something else.

WORKING AROUND THE YARD

General
In this day and age, men and women interchange jobs/ activities much more easily than in the past. The stereotypes of what is exclusively women's and men's work are melting away. In general, though elderly women are more adapted to performing household chores; elderly men would usually prefer to do household repairs and possibly work outdoors. I have had requests from female caregivers to put in a section for more male-oriented activities because their husbands/fathers absolutely refused to help around the house.Persons with Alzheimer's disease/dementia (AD) need to keep active to preserve function and to create some degree of happiness. This does not mean that your day has to be a mad rush, but a few "useful" activities can make your person less restless and easier to care for. A bored person with AD who just sits in a chair all day can be a problem. If this happens, first be sure to have your person evaluated for depression. Then try ideas that you think will suit your person. The ideas listed below, as with the other sections in the workbook, can be used for men or women. It just depends on what makes each individual happy.

Can perform outdoor yard tasks without help
Being able to perform any outdoor yard tasks without help is both a blessing and a curse. For a person diagnosed with dementia, it is only a matter of time before his/her attention wanders. There are many dangerous yard tools that your person may want to continue using. Evaluate carefully, and consistently assess what your person can actually safely do. It would seem to be wise to begin removing tools that may soon pose a danger. You would hate to find out your person can no longer use a tool because s/he had a disabling injury with that tool.

• If your person enjoys working outdoors performing yard activities and still wants to do these activities, hang around and watch to see how well s/he works.

Notes:

• Be alert for lapses in safe behavior.

• Try and be unobtrusive as you allen support and make things safe.

• You can offer to be a helper on whatever project s/he might want to work on, so that you can be useful and supervisory at the same time.

• Work on a parallel project so that you can observe and help if there are problems.

• Keep thinking ahead to what may pose the next danger. Slowly begin removing dangerous items and place easy access to safe ones. Provide materials for projects that only need a rake or shovel verses a lawn mower.

Can perform outdoor yard tasks but needs supervision

If your person cannot safely work in this area by him/herself, volunteer to be a helper. Or if it is your yard project, suggest your person be the helper and assign him/her various simple tasks.

• Start as a helper and transition to the project guide. Make this transition as painless as possible, while still keeping safety in mind.

• Compliment your person on his/her skills as you offer increasing levels of assistance.

• Emphasize the idea of doing projects as a team.

• Observe his/her tool techniques before turning your back on that activity, or you can work next to your person so that you can see what is happening.

• Observe his/her abilities with any tools as they all have dangerous aspects.

• Work with your person to perform the activity rather than going on to another task.

• Give yourself more time to work on any project, so that you can work with your person as a team without pressure.

• Encourage your person to locate tools or items like shovels, trowels and rakes. This will help his/her organizational skills while they still exist. If it takes some time just let him/her do it at his/her pace. Step in if s/he expresses frustration.

• Emphasize the positive in what your person is able to do, not the negative. Do not blame or chide.

• Expect things to get broken, so make sure s/he is not working on something you cannot afford to lose.

• Set up a parallel project to the one you are working on—with cheaper materials.

• Guide your person toward less dangerous work or tasks such as sorting, stacking or prep work.

• Now is the time to truly implement a plan to actually remove dangerous yard tools, or only use them yourself while your person does something easier.

• Lawn mowers and power yard tools pose special problems. One caregiver tells the story of her husband getting out the lawn mower and mowing up the astro-turf on the patio before she could stop him. Lock up the yard tools until you can supervise.

• At this stage do not let your person work in the yard without your

supervision, unless you have created a completely safe area where no accidents can happen.

• Also lock up all yard poisons and fertilizers. It is just like "child-proofing" your house. Your person will not remember how to use these products safely.

• If working in the yard is not your area of expertise or pleasure, but it is for your person then think about finding some family members or friends who can help work with your person a few times per week and implement some of the above ideas. It could be a team contribution to organize all the tools and to create a special place where they can be locked up until used with supervision. What a great gift that could be.

Cannot perform any outdoor yard tasks

By this time your person may not care about yard work anymore. If s/he still seems interested, find very simple tasks that can be done while you work close by. The activities for your person do not have to be real, just simple and seemingly "useful."

• Continue to allow your person to work in the yard as long as s/he is still interested. Continue to position him/her into a safe zone with activities that are safe and appropriate.

• Consider creating a garden and using the ideas in the **GARDENING** section (below) as parallel activities while you perform yard chores.

• Use a potting bench or workbench to allow your person to plant seeds or small plants.

• Simply let your person dig in the dirt, or carry and play with a yard/garden tote.

• Let your person sit at the table and sort seed packets (make sure they are not poisonous in case your person eats them) or any yard related items. This can be made into a repetitive project by dumping sorted items back into the unsorted box to be sorted again—just like laundry.

GARDENING

General

Working outdoors with plants is a soothing and soulful activity. Gardening can bring back special memories of youthful time spent in a family garden. It is also wonderful for you and your person because you can easily work together. You do not have to use any dangerous tools except for a garden tiller, which only needs to be used in the beginning. If you are already a talented gardener, remember to remain flexible. You cannot predict all of your person's behavior. S/he may pick or prune plants unexpectedly. Be prepared for some loss and replant when you can. This activity will be most enjoyable if you keep it light. Do not scold or deride your person for mistakes.

Can perform gardening tasks without help

Being able to perform any gardening task without help is both a blessing and a curse. For a person diagnosed with dementia, it is only a matter of time before his/her attention wanders. Most gardening tools do not pose a problem, except for a tiller. If your person insists on using it and has the ability, still provide supervision. It may be easier to get rid of the tiller and just hire

Notes:

someone to do it. Tell your person it was a "good" deal and that you both need to concentrate on some other aspect of planning or creating the garden.

- If your person enjoys gardening and still wants to do these activities work together and watch how well s/he works.

- Be alert for lapses in safe behavior or distracted behavior.

- Try to be unobtrusive as you offer support and make things safe.

- You can offer to be a helper on whatever project s/he might want to work on, so that you can be useful and supervisory at the same time.

- Work on a parallel project so that you can observe and help if there are problems.

- Keep thinking ahead to what may pose the next danger. Slowly begin removing dangerous items and place easy access to safe ones.

- Try not to plant toxic plants and remove any toxic plants from the garden. You never know when your person might try to nibble on a plant. You can call your local poison control or nursery to check over your list of current plants to identify any that could pose a problem.

Can perform gardening tasks but needs supervision

If your person cannot safely work in this area by him/herself, work with him/her. Gardening can provide pleasant companionship. You can easily take charge of the garden and assign your person various simple tasks. Especially remember to be supportive and keep gardening time as pleasant as possible.

- Start as a helper and transition to the master gardener. Make this transition as painless as possible, while still keeping safety in mind.

- Compliment your person on his/her skills as you offer increasing levels of assistance.

- Emphasize the idea of doing projects as a team.

- Observe his/her tool techniques before turning your back on that activity, or you can work next to your person so that you can see what is happening.

- Observe his/her abilities with any tools as they all can have some problematic aspects.

- Work with your person to perform the activity rather than going on to another task.

- Give yourself more time to work on any project, so that you can work with your person as a team without pressure.

- Encourage your person to locate tools or items like shovels, trowels and rakes. This will help his/her organizational skills while they still exist. If it takes some time, just let him/her do it at his/her pace. Step in if s/he expresses frustration.

- Emphasize the positive in what your person is able to do, not the negative. Do not blame or chide.

- Expect things to get broken, so make sure s/he is not working on something you cannot afford to lose.

- Set up a parallel project to the one you are working on—with cheaper materials.

- Guide your person toward less dangerous work or tasks such as sorting, stacking, prep work or minor pruning (like removing dead blossoms).

• Let your person manage the yard tool caddy—removing and replacing items as needed.

• Buy tools with big handles or ones that can be easily managed by your person.

• At this stage do not let your person work in the garden without your supervision.

• Also lock up all garden poisons and fertilizers. If you need to use these products, place them back in the locked unit as soon as you are finished. It is just like "child- proofing" your house. Your person will not remember how to use these products safely and may actually eat some. Keep the number for poison control handy.

• Watch out for potting soil ingredients like perlite, vermiculite and peat moss because they can get into your person's eyes if s/he is not careful.

• If working in the garden is not your area of expertise or pleasure, but it is for your person; then think about finding some family members or friends who can help work with your person a few times per week and implement some of the above ideas. You could take this time as respite or join in the fun. Gardening is a great stress reliever. You could let the friends or family members manage the garden. You and your person can just come in and play. This could be a great gift for the both of you.

Cannot perform any gardening tasks
By this time your person may not care about the garden work anymore. If s/he still seems interested, find very simple tasks that can be done while you work close by. The activities for your person do not have to be real, just simple and seemingly "useful." Or simply bring your person out to sit in the garden while you work. This can be very soothing for your person.

• Continue to allow your person to work in the garden as long as s/he is still interested. Continue to position him/her into a safe zone with activities that are safe and appropriate.

• Use a potting table or workbench to allow your person to plant seeds or small plants. This can be set up on the patio where you can supervise more easily.

• Simply let your person dig in the dirt, or carry and play with a yard/garden tote.

• Let your person sit at the table and sort seed packets (make sure they are not poisonous in case your person eats them) or any yard related items. This can be made into a repetitive project by dumping sorted items back into the unsorted box to be sorted again—just like laundry.

• Again just let your person come out and sit in the garden.

DRESSING SELF

General
This is a key activity of daily living, and one that is wrapped up in a person's sense of self and function. Allow your person to dress him/herself for as long as possible. This will keep him/her independent and more easily cared for.
Initially his/her clothes may be fashionable or tricky to get on. As you are able, replace these clothes with comfortable clothes for your person. Don't wait until things get very confused and difficult. Phase

Notes:

the new simpler clothes in slowly, and they may not even be noticed. Buy clothes that will be easy to care for and can handle food spills without permanently spotting. Get them in solids rather than stripes or plaids so they will go together more easily. Buy them in the same color family, i.e. tans, or blues or greens etc. again for easy matching. You can purchase clothes that have little difference between the fronts and backs. This can be very useful when getting your person to dress him/herself.

In the same vein, when buying new clothes, think about clothes that use Velcro instead of zippers or buttons. Elastic waists in pants eliminate the need for a belt or suspenders. It's one more item of clothing you can eliminate. Sweaters, pullovers, tee shirts, sweatshirts and sweatpants are easier to manage. If you can afford to, discard clothes, as they get ratty or permanently soiled. Your person needs to look clean and tidy when out in the world. Be supportive when giving dressing directions.

Dresses self appropriately
Allow your person to continue to dress him/herself as completely as possible. Help him/her maintain a neat, clean appearance, so that your person can look and feel good.

- Keep observing your person's management of his/her clothes, dressing, and choices.

- Offer gentle reminders or advice if the combinations get too bizarre or if something is completely on wrong.

- Be prepared for your person to eventually act a bit like a school child and possibly throw an occasional fit because s/he cannot locate a favorite piece of clothing. Offer soothing advice and attempt to find the item or distract him/her to another task.

Tends to put clothes on backwards or mismatches clothing
By 5 years of age most people can dress themselves without supervision. If your person is starting to have some trouble getting clothing straight, s/he has begun regressing in this area. Losses in Alzheimer's disease/dementia (AD) are individual. You will now need to discreetly support your person in this activity while attempting to keep his/her self-esteem intact.

- If your person occasionally puts clothes on backwards, try telling him/her to look for the tags on shirts and mark the tags with a colored marker. Gently remind him/her that the tags go in the back.

- Try laying out the clothes with the fronts or backs up depending on your person's rote technique for putting them on.

- If your person does this consistently, s/he probably needs to be supervised while dressing.

- Gently remind your person if they put on the wrong leg of trousers or the wrong shoe.

- Buy some clothing that has no back or front.

- Buy clothes that are easily matched and "mistake" proof.

Tends to reverse use of clothes (shirts for pants)
- Provide gentle reminders as to where each article belongs.

- Lay out the clothes on the bed. This may trigger early recognition of dressing patterns and prompts, provided by parents years ago.

- Supervise your person's dressing and provide gentle reminders.

- Hand your person the correct item of clothing and have him/her put things on one at a time.

Occasionally cannot find clothes in closet or drawers

- Make sure clothes are separated by type (socks, shirts underwear in different drawers).

- Start by labeling your person's drawers with the contents.

- When a written label doesn't work cut out pictures of the clothing from magazines or advertisements. (Socks, shirts, underwear etc.)

- Remove clothing from your person's closet, and just provide the clothes needed for the day.

Cannot recognize own clothing

- Set out your person's clothing.

- Provide verbal prompts for putting on each piece of clothing as needed.

- Make sure you have gotten to your easy dress clothing, eliminating as many obstacles like fronts vs. backs, zippers, and buttons as possible.

- Let your person remain as independent as possible. You may have to allot more time for dressing so that you are not harried. Your person will pick up on your nerves.

Tends to wear too many clothes

Occasionally a person with Alzheimer's disease/dementia (AD) will just put on their day clothes over his/her pajamas or other nightclothes. S/he has lost a sense of what is appropriate.

- During the summer, make sure winter clothing is put away and not accessible.

- Set out your person's clothing.

- Supervise dressing and undressing activities.

- Provide prompts for each piece of clothing as needed and discourage layering.

- Make sure you have gotten to your easy-dress clothing, eliminating as many obstacles like fronts vs. backs, zippers, and buttons as possible.

- Let your person remain as independent as possible. You may have to allot more time for dressing so that you are not harried. Your person will pick up on your nerves.

- Tolerate some layering if s/he insists and becomes agitated. This issue is not worth a melt down.

- Some things can bring ridicule. Wearing underwear on the head can produce quite a public reaction.

Tends to wear too few clothes

- During the winter, put summer clothes away and vice versa.

- Set out your person's clothing.

- Supervise dressing and undressing activities.

- Provide verbal prompts for each piece of clothing as needed and encourage putting on all the appropriate pieces.

Notes:

Notes:

• Make sure you have gotten to your easy-dress clothing, eliminating as many obstacles like fronts vs. backs, zippers, and buttons as possible.

• Let your person remain as independent as possible. You may have to allot more time for dressing so that you are not harried. Your person will pick up on your nerves.

• Tolerate some cut backs on the amount of clothing (as long as your person does not run the risk of getting hypothermic) if s/he insists and becomes agitated. This issue is not worth a melt down.

Likes to go naked sometimes

At home this can be somewhat disturbing, but in public it can be very embarrassing.

• Try to establish that your person is not overly hot or cold and is confused about how to handle it.

• Allow some periods of nakedness in your home, if you can tolerate it and the temperature is appropriate.

• Provide verbal prompts when your person starts to disrobe. Encourage keeping on all the appropriate pieces.

• If your person wants to take off a shirt or blouse, provide another lighter or heavier one to put on—depending on the need.

• Use the great art of distraction so that this problem does not become a circus performance. Use snack food or house activities to distract.

• Supervise dressing and undressing activities.

• Encourage your person if male to at least wear shorts in the house. If female, encourage a loose housedress or Mumu.

Puts shoes on the wrong feet

• Try marking the shoes left and right—perhaps on the shoestrings. Some people lose left and right fairly fast, so marking may be useless.

• Hand your person the shoes one at a time and prompt him/her to put them on the correct feet.

• Make sure your person is comfortably seated while trying to put on shoes.

• Tube socks have no heel so they can't be turned wrong.

• Transition to slip-on shoes or ones with Velcro, because it won't be long before more of the task is lost.

• Try not to put his/her shoes on for him/her. Let your person do it for him/herself to remain as independent as possible.

Cannot find socks and shoes

• Have a place where your person puts shoes and socks each night.

• Take the shoes and socks each night, and you keep them in a safe place. This way your person can't hide them at night. This also can inhibit wandering outside the house at night.

Cannot tie shoes

• Get shoes with Velcro fasteners, or buy slip-ons shoes.

• Allow your person independence for as long as possible.

Notes:

• If you have a young child, have your child help tie your person's shoes. Make sure this is a happy sharing thing and not demeaning for your person.

• As a last resort or if you don't have time to wait, tie or slip on the shoes for your person. You do not want to discourage the skill, and make your person permanently dependent.

Cannot put shoes on

• Get shoes with Velcro fasteners or slip-ons.

• Allow your person independence in some part of the task for as long as possible. You do not want to discourage the skill, and make your person permanently dependent.

• Tie or slip on the shoes for your person.

• If it's around the house, just use slippers or sock moccasins. Be careful to assess the risk of slipping or falling in this type of foot-covering. You can buy some with non-slip bottoms.

Cannot dress self at all

• You will need to dress your person.

• Tell your person about what you are doing before and during each step of the activity.

• Speak calmly and move slowly so that your person does not misinterpret your movements as threatening.

• Take a short time-out if you meet with a lot of resistance.

• Put on basic shirt and pants so that your person does not get tired and begin to resist.

• See the section on **INSTRUCTIONS** (page 33) for ideas on giving one and two-step commands.

• You will need to get skilled in this to elicit the cooperation of your person for even simple things like "lift your arm".

ACCESSORIES

General

For such small items, accessories can cause a caregiver quite a bit of grief. Women usually love their purses. They fret over them even when they have no cognitive impairment. How often do we all fret over lost car keys, gloves, and wallets? If you add to this the inability to cognitively reason out where you might have left the item, you can imagine how your person must feel, and why s/he might feel panicky about all the things in life slipping away.

Can find purse, wallet, or other valuable items

Throughout your person's life, s/he probably had routines, as most people do for placing his/her valuables in a certain place. This is an excellent habit to maintain or even start while your person can still find his/her special items.

• Create a special place like a basket to place valuables in at the end of the day or after an outing.

• Get your person in the habit of putting his/her purse or wallet in the same place every time.

Notes:

• Don't forget a place for glasses as well. If they are prescription glasses then this will be very important.

• Get duplicates made of all special items, including glasses and keys before there is a crisis of loss.

• Get your person in the habit of carrying very few credit cards and little money in his/her wallet or purse. This way when memory gets intermittent and the items start being lost, you will not have lost too much.

• Make lists now of all credit cards your person insists on carrying so that you can quickly call and cancel them after they are lost.

• Think about switching out credit cards for fakes if your person just likes to carry cards. This will also save you grief if/when they are lost. This is tricky if your person likes to use credit cards.

• Consult the **MONEY** section (page 28) for ideas for helping your person safely manage money and credit or debit cards. Now is the time to make some of these preventative changes.

Often cannot find purse or wallet

As memory loss sets in, there are a few things you can do to ease the stress of lost items.

• Keep up with your system of places to put things even when they are not in use. This will no longer be as useful as it was, but it can still help. It can also help your own memory (that will seem a bit frayed as things are getting lost).

• Tie a bright colored handkerchief on the purse.

• Little key finder gadgets are sold that you could attach to the purse, press the button on a small transmitter or clap your hands, and the device sends out a signal.

• If your person cannot find the purse or wallet because s/he is hiding them (this often happens) make sure there is not much money or valuable IDs in the wallet or purse.

• Start keeping your person's ID in a safe place or with you.

• For men's wallets there are wallets made with belt chains that connect to the wallet. These can be bulky, and your person might resist, but try.

• Get duplicates made of all special items, including glasses and keys before there is a crisis. This includes finding look-a-like purses and wallets with similar contents.

• Keep a special file of prescriptions, etc. so that glasses, etc. can be replaced if lost.

Often cannot find glasses

Glasses are a big problem for persons with Alzheimer's disease/dementia (AD). They are just the right size to be easily hidden or lost. With a fading immediate memory, finding them is tricky.

There are several solutions for lost glasses.

• Create a special place like a basket to place glasses in at the end of the day or after an outing.

• Get your person in the habit of putting glasses in the same place every time.

• Get duplicates made of prescription glasses and keep copies of the prescription in a file or with your local glasses manufacturing firm.

• Get a neck string for the glasses.

• Keep a number of pairs of reading glasses around the house in various baskets so that they are all interchangeable and never lost.

Often cannot find hat and gloves

Like other small stuffable items these are easily lost or misplaced.

• With gloves, get a pair for each coat your person wears, and put a string through the arms that attaches the two gloves together, or use the little suspender clips that are sold for children.

• For hats, stuff the hat into the pocket or sleeve of their jacket each time it is removed.

• Buy extra hats and gloves to be produced when the main items have disappeared.

• If your person insists on hanging on to these items—let him/her keep them. Just be sure you have spares when they are needed.

Worries that accessories have been stolen

Each of us has a mental map of our immediate lives. We know who we live with, what clothes we have and like to wear, what jewelry and tools we have. A person with Alzheimer's disease/dementia (AD) at this stage is forgetting where things are. His/her mental map is starting to deteriorate. His/her life seems to be filled with repeating themes especially disappearing items. This can be very anxiety provoking and can induce blaming or accusations of theft.

• Find places for accessories. If they are valuable, take them and put them in a safe place.

• See the section on **PARANOIA** (page 146) for more ideas.

Often hides personal items

This "squirreling" behavior can be very disruptive to the household, especially if Grandma is hiding her false teeth. Take charge of things that are valuable or necessary to daily life. Put them away in a safe place until needed. With false teeth, or glasses, take charge of them until they are needed then give them to your person to use. Observe your person's behavior. S/he can be in the same room with you and act as if you are not there. Is your person stuffing things into the side of the sofa? Is s/he wadding the false teeth into a napkin and absentmindedly putting them in a drawer or worse yet, the trash?

• Keep track of favorite hiding places.

• Be observant of the trash.

• Check these places routinely for missing items.

• See the section on **OBSESSIONS** (page 143) for more ideas.

Can put on coat and zip or button it

The idea is to keep your person as independent as possible. Anticipate that this may become a problem in the future and adjust his/her clothing now while it may be more comfortable to do so. Try not to wait until your person can no longer manage the more difficult clothing item.

• For the impaired person, avoid coats with small buttons and zippers that are very difficult.

Notes:

• Get your person a coat with a large zipper or Velcro fasteners.

• Capes and cloaks may work as well.

Cannot button or zip coat

• If you haven't done so already, get your person a coat with Velcro fasteners. It may be too late for them to manage a large zipper.

• Capes and cloaks may work as well.

Cannot manage coat

• Allow your person to do as many steps as s/he can manage, even if s/he cannot manage the whole task.

• Help your person to put his/her coat on. Discretely inform friends and family that your person needs assistance with putting on his/her coat so that they are not caught by surprise.

• Smooth over any frustrations by acting matter of fact about the need for help.

Cannot manage any accessories

If you haven't done so already, you have to take complete control of all the elements in your person's daily life. This day will not come quickly but it will eventually come.

• Keep tabs on all necessary accessories.

• Put them in a routine place. It is your memory that needs support at this point. Your person may no longer be able to or care to hide his/her things.

PERSONAL HYGIENE – BATHING

General

Good bathing habits seem such a natural part of a healthy person's life. It seems hard to imagine that this skill could be lost, but as your person regresses s/he will again approach those early years when bathing was sometimes a battle. Often bathing anxiety is associated with the fear of the stimuli such as the rushing water of the shower or other fears like slipping on a wet surface. These feelings of being out of control add to your person's confusion creating a vicious cycle and possibly a catastrophic reaction. An inability or an unwillingness to bathe can have consequences both from a health point of view and from a social point of view; but you can be flexible about how often a person actually needs a bath, the way in which the bathing is done and when bathing is done. Caregivers often feel embarrassed if their person is smelly or dirty, as if they have done a bad job. Your person should not be bathed against his/her will. It is very important to keep bathing time as pleasant as possible and to keep this as an independent skill. Assisting your person with bathing can be quite emotionally loaded because of personal and privacy issues.

Bathes self without assistance

Work toward keeping bathing as a comforting and pleasurable experience as well as preserving your person's bathing skills.

• Discuss bathing with your person and determine how often your person needs to bathe and make it part of a routine. The elderly do not need to bathe as often because of decreased oil and other secretions.

Notes:

The skin is not as elastic. It becomes dry, thin and fragile.

• Set up the bathroom with all the safety features needed. Tub and shower bars are usually needed at some point in Alzheimer's disease/dementia (AD). It is best to get them fitted now while your person can become familiar with them and learn their use.

• Make sure anti-slip strips are in place and check to see if the floor rug is slip-resistant.

• Place shower shampoo and soap dispensers in accessible places where they won't fall and cause your person problems trying to pick them up.

• Make sure the temperature in the bathroom is comfortable and that there are no drafts.

• Watch for any sign that your person is becoming nervous or fearful about showering or tub bathing. Try to identify the reason why, so things don't get out of hand.

Can take sponge bath by self

This is a good alternative when a full tub bath or shower is not convenient.

• Have your person use this skill periodically just to keep it up.

• Figure out your plan for setting up the sink and other supplies so that it is easy for your person to accomplish the task.

• Use this as an alternative if your person becomes fearful of the shower or it is difficult to get him/her in and out of the bathtub.

Can take tub bath by self

• Provide a tub or shower chair with secure backrest and rubber tips if your person has trouble getting down into the tub or gets tired during bathing.

• Turn down the thermostat on the hot water heater to prevent scalds.

• Preset the water if you think your person might get confused as to how to get a good temperature.

• Use temperature devices made to protect children from scalding that install on the faucet spout that will turn the water off if it is coming out too hot.

• Make sure there are non-slip pads on the floor of the tub, and install bathtub grab bars to prevent falls.

• Make sure your person has a towel or bathrobe handy when s/he gets out.

• To keep bathing a pleasurable activity, gently rub moisturizer or emollient into your person's skin following their bath. Use of oils in the bathtub makes it slippery; wait until after the bath to moisturize.

Can take shower by self

• Make sure the floor of the shower has a non-slip surface or pads.

• Install grab bars to hold on to.

• Use a shower seat to counteract fatigue.

• Preset the water temperature if your person gets confused.

• Install a shower sprayer handle that can be controlled by your person. This may help overcome the fear of rushing water.

Notes:

• Place shower shampoo and soap dispensers in accessible places where they won't fall and cause your person problems trying to pick them up.

• Change out glass doors for plastic. This could help prevent cuts if your person falls against them

Often forgets and must be told to bathe

• Set up a bathing routine and post it in your person's room so s/he can keep track.

• Try bathing in the morning rather than the end of the day or vice versa depending on the comfort of your person.

Afraid or refuses to use tub or shower

Perhaps your person has begun to have increasing confusion and perceives bathing to be too complicated. The rushing water may be too stimulating. The temperature of the air may be too cold. Your person may have difficulty expressing what is bothering him/her. Your person may just be embarrassed about having their private parts exposed or feeling helpless about requiring assistance.

• Encourage your person to talk about what s/he is fearful of. Ask simple yes or no questions if there is a communication problem.

• Identify the things that are obstacles to enjoyable bathing and remediate what you can.

• Check your person for sores, bruises or rashes. They may find bathing painful and want to avoid it.

• Simplify all parts of the task.

• Set out all supplies that will be needed.

• Make sure the air and water temperatures are comfortable.

• Stay in close proximity if your person wants to bathe in privacy.

• Install a shower or tub sprayer handle that can be controlled by you or your person. This may help overcome the fear of rushing water.

• Settle for a sponge bath if resistance gets too great and then try again another day. (Who knows, your person may forget what bothered him/her the first time.)

• Be insistent but gentle. Don't make bathing a source of anger or anxiety

Afraid of water

This is possibly a case of regression back to a stage where s/he feared the water as a child, or the stimulus from the water is confusing or uncomfortable.

• If you can persuade your person to take a bath, use only about 3 inches of water in the tub, and make sure the temperature is just right.

• Let your person cover his/her shoulders with a towel for extra warmth if needed.

• Install a tub or shower sprayer handle that can be controlled by you or your person. This may help overcome the fear of rushing water.

• Make bathing as pleasurable as possible.

• Give your person a gentle back massage or rub his/her feet.

• Encourage them to experience the bath as relaxing.

Notes:

• If it is your spouse for whom you are caregiving, try bathing or showering with him or her.

• Assist your person with a sponge bath.

Needs assistance bathing

• Help your person to the extent needed.

• Let your person do as much for him/herself as possible.

• Be mindful of your person's need for privacy. S/he can cover various parts with a towel to minimize exposure.

• Let your person cover his/her shoulders with a towel for extra warmth if needed.

• Install a tub or shower sprayer handle that can be controlled by you or your person.

• Don't leave your person unattended for long. S/he could inadvertently turn on the hot water and forget it, or be too long in the water until it gets cold. Remember that unpleasant experiences can make your person refuse to bathe.

• Allow your person to clean his/her own genitals. This will give him/her a feeling of privacy and control.

Must be given a full bath

• Save your back; get a shower stool that allows your person to sit at chair height in the bathtub.

• Install a tub or shower sprayer handle that can be controlled by you or your person.

• Don't do a full tub bath if you can't physically help your person out of the tub. You can wait until a day when relatives or friends can come over and help, or schedule them to come over once or twice a week to help with your person's bath and transfers in and out of the tub.

• Settle for sponge baths in between.

• Get a book on giving a full bed bath, if that is what you need to do. Bed baths need good organization and planning until you get down the routine.

• At the point you need to be giving a full bed bath because your person is now in a more vegetative or less responsive state, you may want to get a good book on home care in general. (See suggested reading list in the appendix.)

TOOTH BRUSHING

General

One of the big dental health problems with persons with Alzheimer's disease/dementia (AD) is gum disease. Pay attention to your person's dental hygiene. Gum disease or tooth decay has the potential to make both your lives miserable. Make it a habit to look at your person's mouth regularly for redness of the gums, sores, tongue lesions, or tooth decay. A balanced diet and vitamin C supplementation will help deter gum disease. If your person has dentures, check with him/her periodically and ask how the dentures are feeling. Make sure to visit the dentist routinely, and ask about needed care if there are ill-fitting dentures, or painful teeth or gums.

Notes:

Brushes teeth or dentures by self

Stay with this habit. Make it part of a routine.

- Set up all supplies in reasonable order in the bathroom, so that your person does not get angry or confused trying to find something.

- Supervise the task periodically to see that your person is not missing a step or doing anything in the wrong order.

- Give gentle reminders if you notice mistakes. Make sure to support your person's skills.

Needs reminder to brush teeth or dentures

If you incorporate this into a routine, it will only take gentle prompting to get your person to brush.

- Cleaning dentures requires more steps than brushing teeth. There is the potential to lose, or drop the dentures. You may want to supervise this activity more closely than brushing.

- Give gentile reminders each day at the time when tooth brushing is needed.

- Post daily activities (i.e. tooth brushing and bathing) on a list or schedule in your person's room if that helps keep him/her organized.

Occasionally uses wrong cleanser for teeth or dentures

- Remove unnecessary items from the area where tooth-brushing supplies are kept. (Dementia-proofing). Think of it in the same context as "childproofing" a bathroom.

Handles dentist visit well

Good dental health is very important. Enlist your dentist's help with your person. Try to make any trip to the dentist as pleasant as possible.

- Always inform any dentist that your person has AD. More professionals are reading up on how to handle this condition and will react with more sensitivity. They would also need to know to watch the anesthesia carefully in case of some unusual reaction.

- Encourage dental staff to use the same gentle procedures they use on children for your person. This way they will tend to patiently explain what they are going to do each step of the way. Your person will react very poorly if s/he gets confused and paranoid. S/he may begin thinking that all those tools coming at him/her are to cause harm. It does not take too much imagination and empathy to see how the dentist office could make a confused person very frightened.

Doesn't handle dental visit well

This can be a very difficult situation because it does not take too much imagination and empathy to see how the dentist office could make a confused person very frightened. Encourage the dental staff to use the same skills they use with small children, but not to be demeaning. Use the above suggestions plus those below.

- Try to offer your person incentives for a good visit to the dentist.

- Insist that the dental staff patiently explain every step of the procedure as they go along. Do not have them give any long pre-instructions except to you privately. Your person will not be able to retain anything that is said except in the here and now.

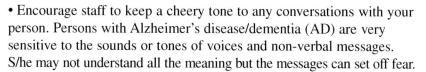

Notes:

• Encourage staff to keep a cheery tone to any conversations with your person. Persons with Alzheimer's disease/dementia (AD) are very sensitive to the sounds or tones of voices and non-verbal messages. S/he may not understand all the meaning but the messages can set off fear.

• Have procedures separated into small segments. Your person may not be able to tolerate long marginally uncomfortable procedures. A tired person with dementia can get cranky and overreact to the situation. If you have to pull out of a procedure in the middle that could be traumatic and possibly painful.

• See if your dentist could get your person in almost as soon as you arrive. Long waits may provoke anxiety.

• Bring comfort items for your person to hang on to if s/he becomes scared.

• Stay with your person through the whole procedure and offer words of encouragement and comfort. Don't hesitate to re-explain in simple terms anything that the dental staff says to your person.

Needs help cleaning dentures

Cleaning dentures with a toothbrush is easier for a person with AD than the more complicated process of using denture-cleaning solutions.

• Set up all the supplies for your person and let him/her scrub away. This could be an easy and satisfying task for your person.

• Supervise periodically or throughout the whole task depending on how much you can trust your person not to eat the cleanser or flush his/her dentures down the toilet.

• Let your person brush the dentures in the kitchen sink while you are working nearby. This might seem odd but could be convenient if you are busy and don't want to stand in the bathroom watching your person brushing.

Cannot recognize toothpaste or toothbrush or needs help with tooth brushing

If your person has lost some ability to recognize the toothpaste or brush, or is forgetting a few of the steps, but can still perform the task, then continue to coach him/her through the task.

• Set up the supplies for the task in an orderly manner. Line up the toothbrush and paste on a light colored towel, where your person can easily see them. See if your person can complete the task with this little bit of support.

• See if your person likes one of those children's (smaller, brighter colored, and easier to use) electrical or battery-operated toothbrushes. If your person using a child's brush bothers you, use an adult one. These brushes do the brushing for you. You may still have to prompt your person to move it around in his/her mouth. Note: If your person does not react well to this, it may be the buzzing sound in his/her head when you put the toothbrush in the mouth or the sensation on the gums.

• Remove other items from the vicinity so there are no other choices like hand soap, shaving cream, or denture cream. You will want to dementia-proof your bathroom at this point because your person may eat something by mistake.

• Place the toothpaste on the brush for your person and prompt him/her to continue the task.

Notes:

• Supervise the full activity so that your person does not eat or swallow the toothpaste (or a bunch of other odd things—use your imagination and your memory if you raised children).

• Gently prompt your person at any point in the task that becomes a stumbling block. Keep as much of it independent as possible.

• Give verbal prompts slowly and one step at a time:

"Get the toothpaste."

"Remove the top."

"Put the toothpaste on the brush."

"Brush teeth."

• Keep plenty of toothbrushes and toothpaste stashed away in case they disappear, especially if your person has a penchant for hiding these items.

• Keep these items out of sight until they are needed again. This is especially important if your person likes to hide them or toss them in the trash.

• Look in the trash before tossing it out to see if any good items have been placed there.

• Let your person brush his/her teeth in the kitchen sink while you are working nearby. This might seem odd, but could be convenient if you are busy and don't want to stand in the bathroom watching your person brushing.

Cannot brush teeth at all

If you need to do the task for your person, do it gently and with patience.

• Explain every step as you are doing it so that your person does not think you are trying to hurt him/her.

• Try using an electric or battery-operated toothbrush. It will take less effort and produce better results. Note: If your person does not react well to this, it may be the buzzing sound in his/her head when you put the toothbrush in the mouth or the sensation on the gums.

• Use less toothpaste so that there is not so much to spit out.

• Gently encourage your person to take sips of water and spit in the sink.

GROOMING

General

If your person is looking well, the public will tend to treat him/her with more respect. This has a positive effect on his/her self-esteem. You will tend to look more favorably on your job as caregiver as well.

Can brush or comb hair by self

Make sure you keep the grooming tools in the same place in your person's room.

Compliment your person on how well s/he looks when finished.

• Keep brushes, combs and other grooming items organized on a dresser top or in a drawer.

• Keep the items simple and small in number to decrease confusion.

• Purchase items with easy-to-use handles and of a larger size.

• Have a mirror at a reasonable height for your person to use for grooming.

Notes:

Can shave self

Depending on how your person shaves, whether with a blade or electric, begin to monitor shaving habits.

- Invest in an electric shaver that is easy to use and requires minimal care. Get your person used to using it (if s/he is not too resistant). Sharp razors could create problems in the future.

- Make certain your person is using his/her shaving tools properly.

- Encourage your person to shave every day. If he has a moustache or beard, help him with keeping it trimmed.

Can apply own make-up

If your person can apply her own makeup, let her. At this stage, observe what she does. Watch as she applies the makeup, what is her routine and how much and where.

- Keep make-up items simple and organized to minimize confusion.

- Provide a magnifying mirror to aid in application.

- Encourage the use of a beauty parlor or spa for waxing eyebrows and other beautifying treatments like pedicures and manicures. If you are a female caregiver, this could be a monthly outing for the two of you (if you can afford it). See if the spa will give your person a senior discount. This can be a great gift from relatives and friends.

Needs help with care for hair

As usual let your person do as much for him/herself as possible.

- Encourage your person to take his/her time in personal grooming.

- Let your person know how attractive s/he is when well groomed.

- Think about dry shampoos, if your person needs to have a quick hair wash.

- If your person is resistant to shower or bath shampooing, try a sink wash with gentle shampoo.

- Try simple styling so that there is little chance of hair getting too mussed up during the day.

- If your person has long hair, and you can talk him/her into getting it cut short, it will save you a great deal of trouble over time.

- Consider a spa, barbershop, or beauty parlor to get a wash and a style. It is so easy for them to position your person's head over the sink. It may also be very soothing for your person to be pampered this way.

Needs help to shave

You will want to keep this as simple as possible for your sake and that of your person.

- Invest in an electric shaver that is easy to use and requires minimal care. Get your person used to using it (if s/he is not too resistant). Sharp razors can create problems. If your person uses an electric razor you can just watch and help him get all the spots.

- Sit with your person as he shaves and supervise.

- Help him break down the task, if he insists on using a blade rather than electric razor:

   "Wash Face."

"Apply Lather."

"Shave and clean the razor after each time."

"Clean and put the razor away"

• Compliment him on a good job

• If your female person with dementia needs help with shaving armpits and legs, consider an electric razor. There is so much danger of cutting your person while trying to help her.

• Consider a spa, barbershop, or beauty parlor to get a get a shave or a wax. It may also be very soothing for your person to be pampered this way. This could be a great gift from a relative or a friend.

• Consider a home beautician. There are a few that will come to your house to cut hair and offer other beauty/grooming services.

Needs help to apply make-up

At this point you may want to simplify matters. Your person can do some very garish things with make up if you aren't careful; and you don't want your person looking clownish.

• Reduce your person's make-up routine as much as possible—perhaps to a bit of foundation and some blush plus lipstick.

• Have your person get used to having you apply any desired eyeliner so that she doesn't poke herself in the eye.

• Allow your person to do as much as possible with gentle prompting.

• Give your person a hand mirror, and you can do her basic makeup. Work slowly and explain what you are going to do before you do it.

• Avoid eyeliner unless you use a simple eye crayon and apply a brief outline.

Uses too much cologne or perfume

You take control of your person's cologne or perfume and apply the right amount. If s/he insists on keeping the perfume, try watering it down.

Must be shaved

At this stage, shaving does not have to be done every day—three or so times a week is fine. You can leave the shaving for outings or if it has been too long between shavings.

• Use an electric razor and get a pre-electric shave facial lubricant; it will make it easier on both of you.

• If your person is resistant to being shaved, take your person to the barbershop and have the barber give him a shave.

Requires total assistance with grooming

Expect to spend a little time with your person doing grooming every day. It's a lot of trouble, but it's worth it. If your person looks bad or smells bad, it will discourage you from caring for your person and other people from interacting with him/her. Persevere. It will pay off. When you're done, have your person look in the mirror, and compliment them on how well they look.

• Establish a routine and maintain at least a basic level of cleanliness and grooming: washing his/ her face and hands, combing or brushing hair, and keeping nails trimmed.

• Consider a spa, barbershop, or beauty parlor to get a get a shave or a

Notes:

wax. It may also be very soothing for your person to be pampered this way. This could be a great gift from a relative or a friend.

• Consider a home beautician. There are a few that will come to your house to cut hair and offer other beauty/grooming services.

TOILET-URINE

General

One of the things you can almost count on is that your person will eventually have problems with toileting. Incontinence can be caused by a number of medical conditions, medications, muscle control issues, dementia, stroke or injuries. Your MD should be consulted for any sudden changes in your person's urinary patterns or health. Toileting is a very private affair and loss of normal skills in this area can be very embarrassing for your person. You will need to be very sensitive in handling these matters. Treat your person, as you would like to be treated.

Can hold urine and toilet self

Independence in toileting is very important because toilet behavior is very private. You will want to encourage your person to exercise his/her skills in this area for as long as possible.

• Try to establish a daily toileting routine. This may discourage worry or obsession over when to go to the bathroom.

• Make sure your person is able to identify the bathroom. Label it with a picture and a word if it is necessary. Strips of reflecting tape and a night-light might help him/her find it at night.

• Install any grab bars that may be needed to make getting on or off the toilet easier. You can use something placed in front of your person (a walker or a tray with books etc.) so that s/he has something to hold onto in case s/he he gets fidgety and wants to get up before s/he is done.

• A raised toilet seat may also be helpful. This equipment can be purchased at a medical supply or hardware store.

• Keep the toilet well stocked with toilet paper so that your person does not get confused and has an accident trying to find what s/he needs.

• Do the same for soap and towels for hand washing.

Has an occasional toileting accident

Occasional accidents can be caused by not remembering body signals or confusion over the location of the bathroom or how to handle the tasks. You will need to be aware of your person's bathroom skills in order to anticipate future problems.

• Don't fuss over the occasional accident. Just go about the business of cleaning up and manage your feelings away from your person.

• Try to identify any underlying problems that may be prompting the sudden loss. Is your person ill, does s/he have a bladder infection, and is s/he too weak to reach the toilet? Consult with your person to understand the cause of any symptoms you find.

• Be aware that dribbling could cause some small accidental urination. This problem often starts in middle age for women. It is called stress incontinence. Women can use small to large pads purchased in the

Notes:

sanitary pad section of the grocery store. There are also larger pads made especially for incontinence. What you use can depend on the size of the problem.

• Try to establish a daily toileting routine. This may discourage worry or obsession over when to go to the bathroom and accidents.

• Make sure your person is able to identify the bathroom. Label it with a picture and a word if it is necessary. Strips of reflecting tape and a night-light might help them find it at night.

• Install any grab bars that may be needed to make getting on or off the toilet easier. Also consider a raised toilet seat. You can use something placed in front of your person (like a walker or a grab bar.) so that s/he has something to hold onto in case s/he he gets fidgety and wants to get up before s/he is done. This equipment can be purchased at a medical supply or hardware store.

• Keep the toilet well stocked with toilet paper so that your person does not get confused and have an accident trying to find what s/he needs.

• Do the same for soap and towels for hand washing.

• Get your person to call for you when s/he needs help.

• Use deodorants around the house and on your person as needed to keep any smell of urine down.

Needs help with toileting

This is a most difficult area because your person is truly starting to lose his/her skills with toileting. It will represent some anguish for him/her and a greater level of burden for you. Even within this area of loss you need to allow as much independence as is reasonable and possible. You will need to be very sensitive in handling these matters. Treat your person, as you would like to be treated.

• Install any grab bars that may be needed to make getting on or off the toilet easier. You can use something placed in front of your person (a walker or a grab bar etc.) so that s/he has something to hold onto in case s/he gets fidgety and wants to get up before s/he is done.

• A raised toilet seat may also be helpful. This equipment can be purchased at a medical supply or hardware store.

• Place a commode or keep a handy urinal on the first floor of your house if you do not have bathrooms on both floors. This way your person won't have to struggle to hold his/her urine while climbing upstairs for an urgent urination.

• Identify times when your person is accustomed to using the bathroom and prompt your person to use the toilet at those times. Discourage toileting at other random times. Your person's bladder and bowels can become accustomed to these times and this will be helpful when s/he is no longer able to understand the physical cues to use the toilet.

Be aware that your person may have not have the ability to recognize the difference between receptacles in the bathroom such as a white toilet, sink, and a waste basket. Your person may not know which to use and may be consumed with frustration or use the wrong one.

• Your person may also try to use other items in your house as a toilet (anything that looks like a pot). If you can catch this behavior, take him/her immediately to the bathroom. Cover or remove the object

Notes:

your person used so s/he will not be prompted to use it again.

• Consider easy-to-remove clothing so that it will be easier for your person to handle that part of toileting.

• Be aware that your person may have trouble finding the words to request to go to the toilet. Observe for nonverbal cues like grabbing pants, rocking, or agitation.

• Supervise some part or the entire toileting task. You will have to decide how much time and when in order to allow some privacy. This can be very difficult in a public restroom with a person of the opposite sex. Look for unisex bathrooms when out in public or plan ahead and bring another person of his/her sex along to help. You may have to get very resourceful and ask strangers to help in a pinch. Try to plan ahead.

• Start with more objective cues and information before giving direct instructions. "We are here to use this toilet." and motion to or touch the physical toilet. You can also flush the toilet to prompt your person's memory to start the task. Motion for them to pull down his/her pants. These short cues may be enough to jump-start toileting.

• Give simple, unhurried, and gentle one-step instructions to prompt any parts of the task that are forgotten. This is good when short cues no longer work.

• Keep some toilet wipes in the bathroom. These are good to use to clean up messy toileting.

• Use incontinence pants (adult briefs) on important occasions if you are worried about accidents. These can be pulled up and down if your person does make it to the toilet. They come in disposable and washable styles with removable liners. You will have to shop around and look at the various kinds on the market. Avoid "diapers" at this point if you can, they may encourage dependence and further incontinence.

• Try to be conscious of smell and keep your person as clean as possible. People react very negatively to the smell of urine or feces on a person. This could be very demeaning for your person. Use deodorants around the house and on your person as needed to keep any smell of urine down.

• Purchase washable chair covers that can be placed over waterproof pads to protect furniture if this is an issue.

Incontinent of urine only at night
This will most likely happen before full incontinence and will probably need to be managed with diapers. This will depend on how much urine your person will have over a night and what will hold it.

• Don't have your person drink fluids right before bed. Night fluids can encourage more urine production and a need to urinate at night. This may be enough for a while to reduce night accidents.

• Install night-lights to make it easier to find the bathroom at night.

• Try using a bedside commode to make it easier for your person to get to the toilet at night. Consider a urinal for men.

• Get your person up once a night to go to the bathroom.

• Experiment with adult briefs verses diapers and see which will keep your person the driest.

Notes:

• Be matter of fact if your person likes to play with his/her urinary briefs or diapers. This may be a shocker for most caregivers and you will have to control your feelings. My mother-in-law used to like to take off her diaper in the morning and hug it with the urine dribbling down her neck. Ask for the diaper/pants and place them in the trash or the wash. You can make a simple statement that the diaper belongs in the trash. You cannot bank on this instruction working, but it is worth a try. Try to supervise morning toileting and dressing activities, if you can; and make sure the diapers are put in the trash and that the trash is emptied regularly.

• Use protective bedding to minimize the need to change full bedding. You can buy disposable pads or rubberized flannel baby sheets. Using a draw sheet (a large sheet doubled with rubberized or disposable pad in between placed across the middle area of the bed and tucked in on either side) may also be helpful for easier bed changes and mattress protection. Look for pads with anti-slip qualities.

• Use deodorants around the house and on your person as needed to keep any smell of urine down.

Incontinent of urine

Urinary incontinence begins before bowel incontinence and is easier to control. You should still have anything that seems like full incontinence checked out by your MD, in case it is due to a temporary condition, like the onset of infection or other illness, side effects of medication, or diabetes. A stroke could also produce incontinence.

• Have men checked out by MD for overflow incontinence caused by an enlarged prostate.

• Use incontinence pants (adult briefs). These can be pulled up and down if your person does make it to the toilet. These pants can feel more independent and manageable than diapers. They come in disposable and washable styles with removable liners. You will have to shop around and look at the various kinds on the market.

• Consider easy-to-remove clothing so that it will be easier for you and your person to handle that part of toileting.

• Be very matter of fact and gentle in your instructions to your person about toilet tasks. This is a very sensitive area. Your person will want some dignity and sense of independence if s/he is still walking and able to perform other activities.

• Don't talk about urine problems and diapers etc. in front of your person as if s/he wasn't even there.

• Don't expect incontinence pants /diapers to hold more than one urination.

• Diapers can be used instead of incontinence pants if it is easier for you. Changing diapers after a BM is easier than incontinence pants, as you will have to take off his/her external pants and shoes. This can be a problem in a public restroom.

• Cleanliness is very important. When changing diapers or incontinence pants, wipe the soiled area clean, wash it thoroughly, and dry gently. You may need to carry toilet wipes if you have to clean up in a public restroom or someone else's house.

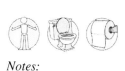
Notes:

• Keep a portable toilet bag (similar to a mother's diaper bag) filled with clean diapers or incontinence pants and wipes, small plastic bags for trash, and even plastic gloves in case the situation is very messy. You can choose from a number of bags that are large and look attractive for this purpose.

• Daily bathing may become more important at this point to reduce the effects of ammonia from urine on the skin.

• Use protective bedding to minimize the need to change full bedding. You can buy disposable pads or rubberized flannel baby sheets. Using a draw sheet (a large sheet doubled with rubberized or disposable pad in between placed across the middle area of the bed and tucked in on either side) may also be helpful for easier bed changes and mattress protection. Look for pads with anti-slip qualities.

• Purchase washable chair covers that can be placed over waterproof pads to protect furniture if this is an issue.

• Use deodorants around the house and on your person as needed to keep any smell of urine down.

• Catheters are a last choice for people with any mobility but may work for bed bound persons. Consult with your MD on the various choices. You may need home health care to instruct you on how to manage a catheter at home.

TOILET- BOWELS

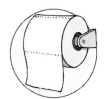

General

Your person cannot be counted on to be able to keep track of when s/he last had a bowel movement (BM). You will need to keep some track of this to detect constipation or diarrhea and get some treatment before things worsen.

 The number of normal bowel movements is very individual and can range from every other day to twice daily. Make note of your person's normal pattern. Occasional accidents can be caused by bowel problems (constipation or diarrhea) or confusion over the location of the bathroom or how to handle the tasks. You will need to be aware of your person's bathroom skills in order to anticipate future problems. Incontinence of bowels usually does not happen until late stages of Alzheimer's disease/dementia (AD) and is usually not reversible. Other unusual things can happen. When your person regresses to a certain point, s/he may become interested in the contents of the toilet. When children are first toilet trained, they often like to look and see what their BMs look like and may reach in and touch them. Your person may also do this and may place their BMs in unusual places (my mother-in-law would put them on the dresser-lined up with her brush and comb).

Can manage own toileting with bowel movements

Independence is very important with toileting. It can make your person feel excessively dependent if s/he cannot manage this activity.

• Make sure your person is able to identify the bathroom. Label it with a word and a picture if it is necessary. Strips of reflecting tape and a night-light might help them find it at night.

• Install any grab bars that may be needed to make getting on or off the toilet easier. Doing this before it is truly needed will make future

Notes:

adjustments easier. Your person does not have to use them if s/he doesn't want to, but you never know when s/he may have a bad day and the bars will come in handy.

• Keep the toilet well stocked with toilet paper so that your person does not get confused and have an accident trying to find what s/he needs.

• Do the same for soap and towels for hand washing.

• Dementia-proof your bathroom early on. If your person is capable of using cleaning products to clean the bathroom, you can furnish what is appropriate at the time. This way there is less risk of poisoning.

• Make note of your person's normal pattern for future reference in tracking constipation or setting up toileting programs.

Has diarrhea

This is not a normal situation. Your person may have food poisoning or a virus. You may need to consult your MD if the condition persists or is in combination with vomiting.

• Give your person whatever medication your MD recommends. There are many over the counter preparations, but it is a good idea to get a recommendation from your MD for any of these preparations. Your MD knows your person's health picture and can give you the gentlest medication and dosing. If you overdo these kinds of medications, you could disturb normal bowel function. It is good to ask for general recommendations well ahead of time, and then you have a first aid plan.

• Keep up with fluid replacement. Dehydration can be a problem for persons with Alzheimer's disease/dementia (AD). Use regular or electrolyte juices (like Gatorade) if your person likes them. Otherwise water is always good as is.

Is prone to constipation

Constipation is slightly more normal than diarrhea but should be handled with good health habits if possible.

• Make note of your person's normal pattern for future reference in tracking constipation or setting up toileting programs.

• Offer plenty of fluids-8 glasses of water or juice (use low sugar juice if sugar is a problem). A glass of warm water in the morning may encourage intestinal motility.

• Offer more fresh fruits and vegetables—cooked or raw. This may have to be adjusted depending on your person's ability to chew.

• Offer more fiber foods: whole grain cereals and breads, bran muffins, dried fruits, brown rice, wheat germ, and nuts.

• Try foods that ferment quickly in the stomach (if tolerated) like sauerkraut, sauerkraut juice, sour dough bread, cabbage juice, tomato juice, or lemon juice in warm water.

• Cut back on processed foods (such as sweets) and high fat foods.

• Encourage daily exercise. Walking is very good for this.

• Try to establish a daily bowel habit. This may discourage worry or obsession over constipation

• Limit intake of antacids. If this is a problem consult with your MD for dosing that does not cause constipation.

Notes:

• Try bulk forming fiber products that can be added to liquid. The pills may be easier to swallow, but there is a tendency not to drink enough water and this may be more aggravating than helpful. Try a number of different types until you find one that works and is easy enough to get down. There are some new clear ones on the market that may be easier to drink. If you are not sure which to choose, consult your MD.

• Consult your MD before you start with strong laxatives. Your person can become dependent on them and this will cause even more problems. Your person may need to start with mild stool softeners first, before any laxatives.

Has an occasional bowel accident

Occasional accidents can be caused by bowel problems (constipation or diarrhea) or confusion over the location of the bathroom or how to handle the tasks. You will need to be aware of your person's bathroom skills in order to anticipate future problems. Incontinence of bowels usually does not happen until late stages of Alzheimer's disease/dementia (AD) and is usually not reversible.

• Encourage daily exercise. Walking is very good for treating or preventing constipation and establishing bowel habits.

• Try to establish a daily bowel habit. This may discourage worry or obsession over when to go to the bathroom.

• Make sure your person is able to identify the bathroom. Label it with a word and a picture if it is necessary. Strips of reflecting tape and a nightlight might help them find it at night.

• Install any grab bars that may be needed to make getting on or off the toilet easier. You can use something placed in front of your person, (a walker or a tray with books etc.) so that s/he has something to hold onto in case s/he gets fidgety and wants to get up before s/he is done.

• Keep the toilet well stocked with toilet paper so that your person does not get confused and have an accident trying to find what s/he needs.

• Do the same for soap and towels for hand washing.

• Use incontinence pants (adult briefs) on important occasions if you are worried about accidents. These can be pulled up and down if your person does make it to the toilet. Avoid "diapers" at this point if you can; they may encourage dependence and further incontinence. This will depend on how your person is doing with bowel continence. They come in disposable and washable styles with removable liners. You will have to shop around and look at the various kinds on the market.

• Try to keep normal habits and routines so as not to make your person excessively dependent.

• Be matter of fact if your person likes to play with his/her BMs. This may be a shocker for most caregivers and you will have to control your feelings. Pick up the BMs and place them back in the toilet. You can make a simple statement that they belong in the toilet. You cannot bank on this instruction working, but it is worth a try. Supervise bowel activities if you can and make sure the BMs are flushed.

Incontinent of bowel movements

Have anything that seems like full incontinence checked out by your MD, in case it is due to fecal impaction (extreme constipation which plugs the

Notes:

bowel but allows some loose stool to pass) or drug side effects. True bowel incontinence should be a late stage problem and at this point other skills will be lost as well. Your person may be very dependent on you at this point.

• Use incontinence pants (adult briefs). These can be pulled up and down if your person does make it to the toilet. These pants can feel more independent and manageable than diapers. They come in disposable and washable styles with removable liners. You will have to shop around and look at the various kinds on the market.

• Be very matter of fact and gentle in your instructions to your person about toilet tasks. This is a very sensitive area. Your person will want some dignity and sense of independence if s/he is still walking and able to perform other activities.

• Don't talk about bowel problems and diapers etc. in front of your person as if s/he wasn't even there.

• Don't expect incontinence pants /diapers to hold more than one BM.

• Diapers can be used instead of incontinence pants if it is easier for you. Changing diapers after a BM is easier than incontinence pants, as you will have to take off his/her external pants and shoes. This can be a problem in a public restroom.

• Cleanliness is very important. When changing diapers or incontinence pants, wipe the soiled area clean, wash it thoroughly, and dry gently. You may need to carry toilet wipes if you have to clean up in a public restroom or someone else's house.

• Keep a portable toilet bag (similar to a mother's diaper bag) filled with clean diapers or incontinence pants and wipes, small plastic bags for trash, and even plastic gloves in case the situation is very messy. You can choose from a number of bags that are large and look attractive for this purpose.

• Daily bathing may become more important at this point to reduce the effects of ammonia from urine and BM matter on the skin of your person.

• Use protective bedding to minimize the need to change full bedding. You can buy disposable pads or rubberized flannel baby sheets. Using a draw sheet (a large sheet doubled with rubberized or disposable pad in between placed across the middle area of the bed and tucked in on either side) may also be helpful for easier bed changes and mattress protection. Look for pads with anti-slip qualities.

SKIN & FOOT CARE

General

This is part of a good grooming program and is especially important for diabetics. Skin does become more fragile and less resilient as we age. Skin begs for good treatment at all ages and stages of life. Consult with your MD about what you can and cannot do for a diabetic if that is an issue.

Can care for own skin and skin is in good shape

• Encourage your person to apply lotion after bathing.

• Try a number of different types, until you find one with the right amount of moisturizer and an acceptable fragrance.

• Apply special ointment to any rashes. Consult with your MD for what kind is appropriate.

Skin dry

• Application of daily lotion helps to minimize this problem, which can be very uncomfortable in the winter.

• Try a number of different types, until you find one with the right amount of moisturizer and an acceptable fragrance.

• Include hands and feet in the lotion regime. Be careful of slippery feet.

Foot and toe nail care are needed

A regular routine of nail care is best. You will have to assist because this can get complicated for most people.

• Set up a weekly routine to do foot and nail care. This may be best done after a bath when nails and skin are soft.

• Examine feet for cuts, ingrown nails, nail fungus, and athlete's foot. Treat any of these as needed. Consult your MD if the conditions do not respond to over the counter medication.

• Obtain good equipment for the job. You may need heavy-duty clippers or nail scissors.

• Consider a pedicure from a spa or nail salon (pretty cheap although you may have a hard time with the odor of the acrylic nails) every once in a while.

Has cuts, wounds, or lesions

• Treat small cuts and abrasions as needed with simple preparations.

• Consult with your MD for all large cuts or ones that heal poorly. Your MD will provide you with a regime to deal with this health problem.

• Examine your person's skin for any cuts or abrasions during bath time, so that they can be treated in a timely manner and will not get infected.

• If your person is bed bound, it is very important to check his/her skin frequently for any breakdown. Especially examine any areas (buttocks, ankles, heels, and elbows) that constantly contact the sheets or could be abraded by movement.

• If bed bound, move the position of your person in the bed every 2 hours to prevent skin breakdown. This is one of the reasons why care of the bed bound person is so exhausting.

• Get some training on how to care for a bed bound person with Alzheimer's disease/dementia (AD).

Cannot perform own skin care

• Maintain good bathing habits. Cleanliness is your first line of defense.

• Apply the needed lotion when your person is no longer able to do it for him/herself. Be sensitive of privacy issues. This may necessitate minimizing the areas of the body that you apply the lotion to. This is especially true of the buttocks and any area near the genitals.

• Gently massage areas where lotion is being applied. This will improve circulation of blood to the skin.

Notes:

SLEEP

General

Sleeplessness can be problem for persons with Alzheimer's disease/dementia (AD). Medications for sleep often leave a person with AD drugged into the next day. Sleep problems can be aggravated by social isolation, boredom, sensory deprivation or overload, and illness (such as asthma, irregular heartbeats, kidney disease or bladder problems. Sundowning syndrome, a form of sleeplessness that leaves your person with AD awake at night as if it were the day. The best treatment is prevention. You have to keep your person occupied enough during the day to be tired at night. Some times you just have to let your person pace at night in his/her own space and hope to get some sleep yourself. You have to think of yourself (health and well being) as well. Sleep deprivation can cause some amazing mental stress. This problem has been identified in doctors who spend days awake while on call as interns and residents. You do not want to be in the position of mental breakdown. Caregiving is too important of a job.

Has no difficulty sleeping

If your person has no difficulty sleeping, count your blessings and enjoy your own sleep.

- Make sure your person's room is at a comfortable temperature.

- Make sure your person has a comfortable amount of covering. This will be individual.

- Make sure the bedroom and bathroom have night-lights to make it easy for your person to get to the bathroom and back to bed.

Has difficulty falling asleep

- Watch caffeine and alcohol intake, as well as smoking, especially in the afternoon or evening.

- Minimize napping during the day.

- Have your person take a soothing bath before bed.

- Have your MD treat any painful conditions such as arthritis that might disturb sleep.

- Play some restful music on a tape or CD player.

- Encourage your person to read a magazine or book in bed. This will often put him/her to sleep. Check on him/her and turn out the lights after s/he has fallen to sleep.

- Encourage him/her to drink some warm milk or tea. Not too much or s/he will have to go to the bathroom and that may cause sleeplessness during the night.

- Drugs should be a last resort because of the side effects and the likelihood they will compound the problem and make your person sleepy the next day.

Paces at night

Pacing is just a bit different from wandering. Many pacers will just stay awake and fidget in their own room. Wanderers will pace the house and go into your room, or worst of all find a way out of the house.

- Make sure your person gets some degree of activity (i.e. exercise,

Notes:

day care, or visits) during the day and does not take long naps that might hinder his/her ability to sleep at night

• Don't provide over-stimulating activities during the evening. Allow your person to wind down.

• Leave a light on in the bathroom so s/he may be less disoriented if up at night.

• Make sure your person goes to the bathroom right before bedtime. A full bladder can make your person restless.

• Provide a commode if getting to the bathroom is too difficult.

• Make sure your person has covers s/he can pull up if the room becomes cold. Quilts may be more comfortable and manageable than tucked-in blankets.

• It may sound cruel, but you can try locking your person into his/her room if s/he has access to a bathroom or is using diapers. An easy way is to reverse a locking door handle on a bedroom door so it locks from the outside. If you do this, be sure you have a safety plan in case of any emergency.

• Leave the TV or radio on low at night for white noise, which some people find restful.

• Try going in and conversing for a short while in a soothing voice.

• Allow your person to pace, fidget, slam, mutter, and bang in his/her room.

• Dementia-proof your person's room so there is not anything dangerous for him/her to get into.

• Use earplugs for yourself, if your room is very close by.

• Lock your door so that your person cannot burst in on you.

Wanders at night (sundowning)

Night wandering can be hard on a family. If you are not sleeping in the same room with your person, s/he can burst into your room and scare you half out of your wits. Or your person can get restless and wander out of your house into the night. As you might imagine this is an extremely dangerous situation. The first time my mother-in-law did this she got a mile or so away and burst into someone else's home. Thank God they did not shoot her and the dogs did not attack her. The police figured out whom she might be related to and returned her within a few hours. She also tore off her Medic Alert bracelet so that was not helpful in this instance. We slept through the whole incident until the police showed up with sirens blaring. Up to this point my mother-in-law had been very nervous about going out at night. After this episode the children called her Houdini Grandma because she had managed to undo the lock. Many of us have to learn the hard way!!

• Make sure your person gets some degree of activity during the day and does not take long naps that might hinder his/her ability to sleep at night.

• Leave a light on in the bathroom so s/he may be less disoriented if up at night.

• Make sure you lock and /or alarm the doors to the outside so s/he can't get them open. Even install safety locks that will be difficult for a memory-impaired person to learn (often dead bolts that are high out of the visual field). Lock or gate doorways to staircases and other dangerous places.

Notes:

• Dementia-proof your home so there are fewer things to get into.

• If your feel you must go into your person's room to calm him/her down, don't get into an argument. Talk calmly and try to get him/her to go back to bed. If that doesn't work make sure your person is safe; then leave, and go back to bed.

• Take away his/her shoes at night and keep them in your room. This can discourage those with delicate feet from sneaking outside at night.

• Hide the clothes that s/he might be tempted to wear to get up for the day.

• It may sound cruel, but you can try locking your person into their room if s/he has access to a bathroom or is using diapers. An easy way is to reverse a locking door handle on a bedroom door so it locks from the outside. If you do this be sure you have a safety plan in case of any emergency.

• If additional locks are impractical, consider a door alarm that will alert you if the front door or even your person's bedroom door opens (now sold at dementia stores on-line).

• Install a motion detector light outside the house. It will alert you if your person "escapes" in the dark and, it may keep him/her from falling.

SEX

General

Sex is an interesting subject for persons with Alzheimer's disease/dementia (AD) and their spousal caregivers. If you are a familial caregiver, your person is probably celibate and may have been that way for a long time. Your person could be widowed or divorced and sexuality could have fallen to the bottom of the list of his/her needs. In these cases sex may never come up as an interest or a problem.

Has interest in sex and can perform intercourse

Persons with AD tend to have fewer social inhibitions than their well counterparts. Sex is a part of most people's lives and sexual activity in a person with dementia is no exception. We all crave intimacy. People with AD have fewer barriers to satisfying that craving. Persons who were shy may become forward. Persons who have been sexual all their lives may become sexually inappropriate.

• Identify the sexual desires of your person. Are they directed at you the spousal caregiver? If so then you may want to carry on your sexual relationship at a level that has been comfortable for the both of you.

• Be aware of your person's ability to be a willing participant in a consensual relationship. You don't want to be forcing your person to have sex or to perform sex acts that clearly confuse him/her.

Has a sexual partner and desires intercourse

Persons with AD who live with a caregiving spouse may show an increased interest in sex and if the spouse is willing, it could become a nice respite in the caregiving routine. Since a person with AD is likely to be in a state where s/he can't really do much for others, sex may be one place where s/he can give something back. On the other hand, many caregivers may resent being the spouse caregiver (and what amounts to being a parent to

Notes:

their spouse) and a lover. The person they once loved and shared intimate emotions with may be lost to them and as such, the caregiver may feel little attraction to the afflicted spouse. You will have to seek understanding of your own feelings and emotions on the subject. Consult your MD or counselor for help and support.

- Participate in sexual activity with your spousal person at the level, which is comfortable for the both of you.

- If your spouse can perform intercourse and it is pleasing to you, you should continue. If you've had a satisfying sexual relationship during your marriage, it's reasonable to continue if possible. Don't forget good lubricants! Women don't lubricate as readily when older, and the skin on and around men's penises tends to get thinner. Pain and sex don't mix.

- Don't rush it. The intimacy you achieve with your partner helps reinforce the bond of trust and communications on other issues in the relationship. Don't expect your demented person to be too involved with satisfying you. For women, be prepared to bring yourself to orgasm, as he is likely to lose interest after his orgasm.

- Be aware of non-verbal requests for sex. People with very little verbal skills may still motion to their genitals or yours as a request for sex.

- Seek counseling if you feel conflicted about your sexual relationship with your person with. This is a may be a difficult one to sort out for yourself.

Does not have a sexual partner and desires intercourse

If a person is healthy, it is normal to have sexual feelings and urges. If you are not a spousal caregiver and your person has no outlet for his/her sexual desires, this could be a problem and lead to inappropriate sexual advances.

- Masturbation is normal and healthy—just not in public. If your person likes to masturbate (or you perceive that he/she does) let that be a healthy sexual outlet.

- Try not to show shock or surprise if you catch your person in the act of masturbation. Just withdraw and allow him/her some privacy.

Cannot perform intercourse but wants to

There are many reasons why a man may not be able to achieve an erection: poor circulation, drugs, and other physical conditions or just being distracted. Consult your MD for help and support with this issue.

- If your spousal person craves intimacy, help them with that. Cuddle, massage, rubbing and touching can go a long way to satisfying these needs.

- See if mutual masturbation works.

- Drugs for enhancing erections may be a marginal idea because of the prolonged effects and the confusion that they may cause. Talk to your MD about what may be appropriate.

Makes inappropriate sexual advances in non-spousal situations

This is a difficult problem for spousal and non-spousal caregivers. You will need to direct your person's energy elsewhere.

- A person with Alzheimer's disease/dementia (AD) may misinterpret inadvertent touching of sensitive areas while s/he is being dressed or bathed. Simply distract your person if s/he makes sexual advances during these activities. Tell your person not to touch you there or distract

Notes:

him/her by asking him/her to hold the shampoo while you are giving the bath. If he gets an erection, just ignore it and work around it. If your person starts touching him/herself, give him/her something to do with his/her hands.

• If your person disrobes or displays sexually inappropriate behavior in front of children, remove your person from the situation and explain to the child in non-sexual terms that your person was unaware of what s/he was doing.

• Do the same if your person does this in public or with friends or relatives.

Makes inappropriate social advances

For children of your person with AD who are the caregiver, your person can relate to you in shocking ways. A mother might relate to a son as if he is the husband. A father may make sexual advances on a daughter who may look like her mother. This would be due to confusion over the ages of people and their features being similar to the people your person has known over his/her life.

• Try gently reminding your person that you are not your person s/he thinks you are.

• Don't argue. This will only make things worse.

• Try to deflect the advances with matter of fact responses and distractions.

• If the advances are harmless, play along for a bit in the drama and then slowly change the subject and/or distract.

Removes clothing inappropriately and tends to expose self

Often, removing clothing has much less to do with sexuality than discomfort or confusion. Your person may find some of the clothes don't fit; they're too warm or too cold. Your person may not recognize the clothing that s/he is wearing and want to be rid of it. This behavior may also be the result of your person having to go to the toilet and forgetting what comes next.

• Get pants that pull on with elastic wastes like sweat pants and pull over shirts. This will discourage the behavior.

• If your person appears half clothed near the bathroom or expresses in some way that toileting is a problem, gently guide him/her to the toilet.

Masturbates publicly

Crotch rubbing, though it can be interpreted as masturbation, may only be your person's indication that s/he has to go to the toilet. If you are a spousal caregiver, this could be a request for sex.

• If your person is doing this, gently encourage him/her to get up and go to the toilet. In either case your person can take care of his/her needs outside of public view.

• If your person is actually masturbating, find some quick distraction and move away from public view.

• Keep a few favorite distraction items with you for this kind of need.

Has no interest in sex

Depending on your person's background this can be a normal state. This is probably due to the process of increasing dementia. If you are in a spousal relationship this may be the end of any sexual responses. If this is a sudden change, see your MD.

Notes:

• Don't make unwanted spousal advances on your person if s/he shows no interest.

• If this is a problem for you, then masturbation may have to suffice.

• If you have difficulty with your feelings of loss or anger around this issue, seek counseling.

EMOTIONAL
FEELINGS

General

Alzheimer's disease/dementia (AD) affects the limbic system of the brain or the emotion center. This damage has the effect of taking the controls off of your person's emotions. S/he can range emotionally from over-to-under reactive as well as responding inappropriately to certain situations. As with all things in AD, how your person will react will be individual, but you will get to know his/her general patterns. Be prepared for surprises and try not to over react. Calm and firm responses will work the best. How your person behaves does not reflect on you as a person. The disease is the cause of his/her unraveling world and brain. The reactions, including the inappropriate ones, are just coping mechanisms to exert some control in a confusing world.

Able to express appropriate emotions

This is a pleasant stage but may still be marked with occasional bouts of normal anger over being dealt a disease like AD. Your person may be aware at this stage of what the disease will do to him/her and be already mourning the loss of normal feelings and abilities.

• Openly listen to what your person has to say about the disease and his/her feelings.

• Consider some levels of anger and sadness normal. If either emotion seems to threaten to overwhelm your person, seek some counseling. This may be the best time for your person to begin to work through the grief process and counseling could be very helpful. Later when his/her cognitive abilities are declining, s/he may not be able to hold his/her thinking processes together long enough to make constructive shifts in emotional approaches to crisis and loss.

• Clarify any wishes your person may have about visiting people s/he loves while s/he is still able to respond appropriately. Reassure your person that relationships with friends and relatives will not be excluded in the future just because s/he has lost some skills and abilities. Your person should not feel his/her life is over. Reassure your person that you will not let others treat him/her as if it were.

• Remind your person that you will try to let him/her live as well as possible while having AD. It is frightening to slowly lose your mental, physical, emotional, and social abilities but there are things your person will be able to do to keep a semi-normal life.

• If your person is in the early stages of AD, encourage him/her to join the Alzheimer's Associations early stage support group in your local area.

• Love does not stop with the diagnosis of AD.

Notes:

Has to be pushed to talk about feelings

At some point your person will become more self involved and less willing to discuss feelings. His/her emotions may be freewheeling, but s/he no longer possesses the ability to exert controls and to intellectualize on emotional issues. This is a normal course for dementia.

- Encourage the expression of thought around emotion if able.

- Openly listen to what your person has to say about the disease and his/her feelings.

- Go with the flow and work with the moments as they present themselves. If your person doesn't want to talk about his/her feelings, that is okay, unless s/he is emotionally out of control.

- Seek some psychiatric or counseling help if your person seems out of control. Medication may be appropriate at that point.

Expresses inappropriate emotions

This is to be expected and coped with for all the reasons stated above. Don't take what is said to be personal. Your person may be your parent but s/he no longer has physical control over you. If you give them emotional control, you will be headed for problems. You are in control now. Your person's expression of emotion is an attempt to communicate, just like children resort to tears and temper tantrums when they don't feel they can get what they want or feel they need. Your person may be in the same predicament. Words and concepts may have failed and emotion is the only form of expression s/he has left.

- React calmly and firmly. Persons with dementia will often mimic what they hear or see in the environment. Your calm voice will help calm him/her down.

- Try to get your person calm enough to express the problem.

- Try to identify what is upsetting your person. Is it something s/he wants but can't find the words?

- Offer reassurance that you can help.

- Try a yes and no approach to ideas that you think may be the source of the problem. "Are you hungry, thirsty, etc?" Picture communication cards may work when your person is too flustered to speak.

- Check for issues in the environment (i.e. feeling too hot or too cold), things you may be doing, problems with your person like hunger, thirst, or need to go to the bathroom.

- Set firm but kind limits for minor temper tantrums. Offer calm reassurance but stay out of the range of fire, in case s/he tries to take a swat at you. If the tantrum goes much farther, it will most likely move to a catastrophic reaction. Read the next section on **MOODS**.

Has catastrophic reactions

This is a big potential problem in persons with Alzheimer's disease/dementia (AD) and one that is best avoided. A catastrophic reaction is a level of agitation that is indicative of total system overload and cognitive breakdown. It can appear as a raging temper tantrum, uncontrollable crying, excessive worry, tension or stubbornness. The environment may just be too stimulating and your person may not be able to self-calm or remove him/herself from the source. Your person can be emotionally impulsive as well as responsive with too much of the inappropriate emotion. Your person is unable at this point to have any empathy for how you or anyone else will react to his/her outburst. Remember this is not a choice but an out-of-control reaction. Do not blame your person for his/her responses. Just deal calmly with the problem.

- Reduce the amount of stimuli in the environment (i.e. noise, confusion, questioning).

- Speak in a calm reassuring voice. Let him/her feel the security that you are in charge and can help.

- Do not restrain or put your hands on your person while s/he is raging. This may escalate the reaction.

- Give simple one-step instructions.

- Give simple explanations as to your approach and actions as you try to help him/her.

- Gently use distraction to move your person away from obsession over the problem.

- Offer gentle touch—stroking an arm, holding hands, and rocking to calm your upset person with dementia when s/he becomes approachable.

- Prevention is the name of the game. Try to problem solve with your person and meet his/her needs before s/he reacts catastrophically.

- Keep notes on problems or issues that tend to provoke catastrophic reactions. Let other care providers know about these sensitive points.

NOTE: List things that cause catastrophic reactions in my person with dementia.

MOODS

The damage to the limbic system in AD causes swings of emotion and mood. For some persons with dementia, the AD does not faze them; they uphold a positive attitude. Others sink into depression and apathy. There is no way to predict, but your person's base line personality can be somewhat predictive. If s/he was a fairly happy person most of his/her life, then the mood may stay in the same range and vice versa. This is only a possible predictor; there still can be sudden swings that make no sense at all. In fact, it is better to be prepared for change and then be pleasantly surprised when negative swings do not happen.

Happy and cheerful most of the time
This would be the best outcome for you and your person, but you cannot count on it. If this is your starting point in caregiving, be thankful. Caring for someone who is pleasant and cooperative has a large effect on your sense of burden.

- Set up your recreational program while your person is cooperative and willing to try new things.

- Identify favorites. This should be much easier with a cheerful person.

- Set up your household routines and chores for your person. It is easier to back off if things get into gloomy periods later in the disease.

- Set up spiritual sessions if you are interested in doing so. This is a great time to identify future wishes (Work with the 5 wishes document – see the Spiritual section) and establish a spiritual signature.

Notes:

Occasionally sad or depressed

This is not entirely unusual, but be watchful and do not take a "just get over it" attitude. Your person may just swing cheery again, but if s/he is actually in the throws of a clinical depression it will not simply disappear. At the point where you become worried that a negative mood is beginning to overwhelm your person ask your MD for an evaluation.

• Roll with your person's moods as they change-increasing and decreasing activities as needed.

• Try gentle persuasion to encourage your person to participate in favorite activities that have a potential to lift the mood.

• Don't argue. It will worsen the problem.

• Seek help from your MD if the negative or depressed mood seems to have a strong hold on your person.

Often sad, depressed, or cries a lot.

A persistent depressed mood is a symptom of clinical depression. At this point your person may not be able to "snap" out of it. It is important to seek an evaluation with your MD because this type of depression responds well to treatment. There is no reason to prolong suffering for both of you.

• Be aware of the signs and symptoms of depression (some of which mimic Alzheimer's disease/dementia (AD):

*Inability to concentrate or make decisions

*Loss of interest in eating

*Lack of interest in being with other people

*Lack of sex drive

*Sadness or crying spells for no reason

*Sleeplessness or excessive sleeping during the day

*Feels excessively tired

*Brooding on unhappiness or worry

*Feeling unwanted or worthless

*Expression that life is not worth living

• Roll with your person's moods as they change-increasing and decreasing activities as needed.

• Try gentle persuasion to participate in favorite activities that have a potential to lift the mood.

• Don't argue. It will worsen the problem.

• Speak in a calm reassuring voice. Let him/her feel the security that you are in charge and can help.

• Gently use distraction to move your person away from obsession over the problem.

• Offer gentle touch—stroking an arm, holding hands, and rocking to calm your upset person with AD if s/he is approachable.

• Seek help from your MD if the negative or depressed mood seems to have a strong hold on your person.

Expresses loneliness and depression

It is wonderful if your person wants to express his/her sad feelings. You will be able to get a better sense of whether this is a temporary bit of nostalgia

Notes:

and self-pity or if it is deep abiding sadness that may indicate clinical depression. Depression is the most common mood problem associated with Alzheimer's disease/dementia (AD).

- Openly listen to what your person has to say about the disease and his/her feelings.

- Go with the flow and work with the moments as they present themselves. If your person doesn't talk about his/her feelings, that is okay, unless your person is emotionally out of control.

- Speak in a calm reassuring voice. Let him/her feel the security that you are in charge and can help.

- Gently use distraction to move your person away from obsession over the problem.

- Offer gentle touch—stroking an arm, holding hands, and rocking to calm your upset person with AD if s/he is approachable.

- Seek some MD/ psychiatric help if your person seems out of control. Medication may be appropriate at that point.

Talks about self-destruction/suicide

This is a serious state of affairs and may be indicative of a person's desire to commit suicide. In general your person's impairments may prevent him/her from being able to carry out a suicide plan, but don't risk it. Talk to your MD as soon as your person begins to talk repetitively about any forms of self-destruction.

- Be watchful for stronger evidence of suicidal tendencies:

  *A plan with identified means ("I want to die, I can take all those sleeping pills you have in the cabinet.")

  *Access to the means (You can remove any means you know of – pills, guns, alcohol, sharp objects)

  *The mention of a time

- Get medical help as soon as you can.

- Prevention is the best medicine. Talk with your MD about your person's depressed moods before it gets to any talk of self-destruction.

Is mostly non-responsive

This is a late stage pattern and moods at this point will be fairly irrelevant. Although you may be able to discern subtle changes in his/her non-verbal patterns, if you have been with him/her long enough.

- Work with the suggestions mentioned above to keep your person in a positive frame of mind. You may have to rely on very subtle feedback to know what is going on.

- Continue to treat your person as if s/he was still responding and giving you positive feedback. You do not know what is going on in the mind of a non-responsive person. Love is always the best way.

- Carry on one-sided conversations if you like. Your person with dementia may enjoy the gentle banter even if s/he cannot respond. S/he will most likely follow the sense of connection, just like a baby.

Notes:

MENTAL ILLNESS
General

It would be a fairly rare problem, for your person to be diagnosed with Alzheimer's disease/dementia (AD) and be actively mentally ill (i.e. manic-depressive, or schizophrenic). If this happens, you may well not be able to handle this in your home as a sole caregiver. Obtain professional counsel if you are thinking of working with this dual diagnosis. Medication and psychiatric support will probably be needed. Nothing is impossible, but you will want to have the best advice before making your decision.

Has clinical mental illness
Which one:

As stated above this is a difficult issue. The symptoms of AD can compound the symptoms of the underlying mental illness or the opposite could occur. The human mind is a vast place and although we know a great deal we cannot always explain what happens with every certainty.

• Work with the professionals involved in your person's life to make a positive health and life plan for him/her. This plan may or may not include you providing sole care in your home.

• Work very closely with these professionals in the event you do decide to take on this challenge.

Has clinical depression
Clinical depression is a challenge but not to the degree of other mental illnesses. Depression is the most common mood problem associated with AD. How successfully you will be able to manage may depend on the treatment your person has for his/her clinical depression. If it is well managed by a team of mental health professionals, you will probably do very well.

• Be aware of the signs and symptoms of depression (some of which mimic AD):

*Inability to concentrate or make decisions

*Loss of interest in eating

*Lack of interest in being with other people.

*Lack of sex drive

*Sadness or crying spells for no reason

*Sleeplessness or excessive sleeping during the day.

*Feels excessively tired

*Brooding on unhappiness or worry

*Feeling unwanted or worthless

*Expression that life is not worth living

• Roll with your person's moods as they change—increasing and decreasing activities as needed.

• Try gentle persuasion to participate in favorite activities that have a potential to lift the mood.

• Speak in a calm reassuring voice. Let him/her feel the security that you are in charge and can help.

• Gently use distraction to move your person away from obsession

Notes:

over a problem/situation.

• Offer gentle touch—stroking an arm, holding hands, and rocking to calm your upset person with dementia if s/he is approachable.

• Seek some MD/ psychiatric help if your person's depression seems to be worsening. A medication adjustment may be needed.

Has talked about suicide

This is a serious state of affairs and may be indicative of a person's desire to commit suicide. If your person has a mental health problem in addition to Alzheimer's disease/dementia (AD), you will need to pay special attention if your person talks of suicide. In general your person's impairments may prevent him/her from being able to carry out a suicide plan, but don't risk it. Talk to your MD as soon as your person begins to talk repetitively about any forms of self-destruction.

• Be watchful for stronger evidence of suicidal tendencies:

*A plan with identified means ("I want to die, I can take all those sleeping pills you have in the cabinet.")

*Access to the means (You can remove any means you know of — pills, guns, alcohol, sharp objects.)

*The mention of a time

• Get medical help as soon as you can.

• Prevention is the best medicine. Talk with your MD about your person's depressed moods before it gets to any talk of self-destruction.

ANGER

General
Anger is one of the emotions that can go out of control from the damage to the limbic system caused by AD. It can range from petty annoyance to full-blown catastrophic or violent reactions. The pattern for your person will be individual.

Expresses appropriate anger
This is within the normal range and may possibly be dealt with by calm expressions of understanding and empathy. Prevention is very important at this stage. You want your person to feel that s/he has been heard, and there is hope things will be different. You want to prevent a full-blown case of rage that will be difficult to de-escalate.

• Openly listen to what your person has to say about the disease and his/her feelings.

• Let him/her vent for a while and don't argue or offer strongly conflicting points of view. This is not about intellect and logic but about feeling.

• Consider some levels of anger and sadness normal. If either emotion seems to threaten to overwhelm your person, seek some counseling.

• Offer some calm resolutions to the problem.

• See if you can find one that will alleviate your person's anger.

Expresses inappropriate anger
This is to be expected and coped with for all the reasons stated above.

Notes:

Don't take what is said to be personal. Your person may be your parent, but s/he no longer has physical control over you. If you give your person emotional control you will be headed for problems. You are in control now. Your person's expression of anger is an attempt to communicate, just like children resort to tears and temper tantrums when they don't feel they can get what they want or feel they need. Your person may be in the same predicament. Words and concepts may have failed, and anger may be the only form of expression s/he has left at that moment.

- Don't take what is said to be personal.

- Don't engage the fight or argue

- React calmly and firmly. Persons with dementia will often mimic what they hear or see in the environment. Your calm voice will help calm him/her down.

- Try to get your person calm enough to express the problem.

- Try to identify what is upsetting your person. Is it something s/he wants but can't find the words?

- Offer reassurance that you can help.

- Try a yes and no approach to ideas that you think may be the source of the problem. "Are you hungry, thirsty, etc.?" Picture communication cards may work when your person is too flustered to speak.

- Check for issues in the environment (i.e. feeling too hot or too cold), things you may be doing, problems with your person like hunger, thirst, or need to go to the bathroom.

- Set firm but kind limits for minor temper tantrums. Offer calm reassurance but stay out of the range of fire, in case s/he tries to take a swat at you. If the tantrum goes much farther it will most likely move into a catastrophic reaction. Read the section: **Has catastrophic reactions** (page 134)

Has violent reactions

An out of control person with AD can be dangerous for you to handle. Again prevention is your first line of defense. Try to de-escalate all arguments or tantrums before they can get catastrophic and or violent. Luckily violent behavior patterns do not happen very often and the behavior can be over quickly. They can be started because of inability to understand what is going on, over-stimulation or by frustrations over inability to do something (i.e. trouble getting dressed, turning on or off a faucet). To you this may be a simple matter, but to your person it may be the last straw. Your person may bite, spit, hit, punch, kick, or throw things. Try to handle the situation with the ideas listed below:

NOTE: If you cannot control the situation and fear for your own safety, call an ambulance, not the police. Ambulance personnel are better equipped to handle a sick elderly person than the police. There is the very nasty problem of "police assisted suicide." Where the police shoot and kill elderly persons who are brandishing perceived weapons (This happened to my father-in-law). In the current legal climate courts are tending to support this type of brutality as defensible. You do not want yourself or your person in this situation. If the police do respond, do not over dramatize the situation. If they think you are afraid of your person harming you, they will be more inclined to react with violence themselves.

Notes:

• You can try ignoring a non-violent screaming fit for a while to see if s/he will quiet down on his/her own.

• Protect yourself. Stay out of the way of the physical violence.

• Use a pillow or other barrier as a shield until the fit of rage is over.

• Do not argue or try to reason with your person. It will only aggravate the situation.

• Try directing your person into a corner away from doors or windows. You don't want a broken window and cuts on either of you. You also don't want your person to leave the house and run into the street in a fit of rage.

• Try silence while you are working with your person. This may help to de-escalate the situation.

• Speak in a calm reassuring voice or a monotone once your person is in a secure place. Your person will respond to the calmness in your voice.

• Try not to communicate fear or anxiety. Let him/her realize there is no threat.

• Let him/her know you are there to help and s/he doesn't have to be upset anymore.

• Offer a future distraction or comfort like tea, or milk and cake.

WORRY, RESTLESSNESS AND AGITATION

General

There is a tendency for persons with Alzheimer's disease/ dementia (AD) to become overly worried, restless, or agitated. Logically a fading memory and inability to carefully reason problems and solutions contribute to this problem. People without cognition problems can sometimes find themselves in a neurotic pattern where a concept plays over and over in their brain seemingly without benefit of rational interruption. This is very much how a person with dementia feels when s/he gets stuck on a problem. To add to it, the problem that is the source of the fretting can be totally false. Ideas fly in and out of your person's head. It is difficult to sort out the truth from the fiction. One time I found my mother-in-law worrying over the supposed death of her sister and why hadn't she been invited to the funeral. You can't argue, but you can gently propose the truth and see what happens.

Occasionally fretful and worried

The idea is to gently move your person from a place of worry and agitation to a more peaceful state. It will often take some creativity and patience to do this.

• Openly listen to what your person has to say about his/her worries and concerns.

• Try to identify what is upsetting your person. Is it something s/he wants but can't find the words to describe?

• Offer reassurance that you can help.

• Don't argue about the truth. S/he may not be able to follow your logic and reason his/her way out.

• Go with the flow of his/her thoughts and feelings offering small reassurances. "It must be a mistake—the invitation was lost in the mail."

Notes:

• Use your creativity in working the conversation to a less worrisome place.

• If communication is a problem, try a yes and no approach to ideas that you think may be the source of the problem. "Are you hungry, thirsty, etc.?" Picture communication cards may work when your person is too flustered to speak.

• Check for issues in the environment (i.e. feeling too hot or too cold), things you may be doing, problems with your person like hunger, thirst, or need to go to the bathroom.

• Distract, Distract, and Distract! If your patience is running out or nothing else works, try offers of cookies and milk or entertainment. Problems can be dealt with later.

Excessively fretful, worried and restless

As agitation increases, you run the risk of a catastrophic reaction. So de-escalation is important but try not to get agitated yourself. Your person can read your emotions and feed off of your frenetic energy.

• React calmly and firmly. Persons with dementia will often mimic what they hear or see in the environment. Your calm voice will help calm him/her down.

• Try to get your person calm enough to express the problem.

• Try to identify what is upsetting your person. Is it something s/he wants but can't find the words to describe?

• Offer reassurance that you can help.

• Don't argue about the truth. S/he may not be able to follow your logic and reason his/her way out.

• Go with the flow of his/her thoughts and feelings offering small reassurances. "It must be a mistake—the invitation was lost in the mail."

• Use your creativity in working the conversation to a less worrisome place.

• If communication is a problem, try a yes and no approach to ideas that you think may be the source of the problem. "Are you hungry, thirsty, etc.?" Picture communication cards may work when your person is too flustered to speak.

• Check for issues in the environment (i.e. feeling too hot or too cold), things you may be doing, problems with your person like hunger, thirst, or need to go to the bathroom.

• Distract, Distract, and Distract! If your patience is running out or nothing else works try offers of cookies and milk or entertainment. Problems can be dealt with later.

Paces a lot

This is just an offshoot of the above problems but it involves a more physical response. Pacing is an attempt to self-calm. Quieting our own minds and bodies is something we initially learn as infants. As the Alzheimer's disease/dementia (AD) progresses your person will resort to more early strategies for self-calming. Simple contemplation can get him/her into trouble. Rocking and pacing can seem to get things going in the right direction.

• Let your person pace some if it doesn't seem to be bothering you or anyone else.

• Distract your person with another activity or a food treat to break the spell of pacing.

Notes:

• Use the above suggestions for calming a session of worry or agitation.

Difficulty with self-quieting

If your person has quite a bit of trouble calming, you will want to try all the above suggestions as well as these.

• Collect a number of favorite objects that seem to calm your person down. This is similar to the techniques you use with small children. What ever works is what matters not how childlike your person is behaving.

• Try stuffed animals. They have been shown through research to promote comfort and calming in persons with AD. Try a few until you find one or two that work. Don't depend on just one—in case it gets lost.

• Try small soft blankets. You can cut off a piece of a larger one if your person likes the sensual feel of it.

• Try other dolls or stuffed toys. If you know of a favorite your person had as a child, try to find one like it. Stuffed monkeys were popular early in the 1900s.

• Creativity is the name of the game. Use your intuition and imagination.

OBSESSIONS

General

The subject of obsessions has been included because it is a problem in AD. It is a similar issue to the worry and agitation mentioned above. In this case your person with dementia will get focused on certain items verses concepts or thoughts. So much has been taken way from your person: his/her memory, intellect, physical coordination, and some of the meaning of their social life. Physical items can be a substitute for these things and may need to be saved at all cost. Don't underestimate the meaning these obsessions may have for your person. Try to be perceptive and work sensitively with each situation.

Has no obsessions

Not too much to do at his point as this is a normal state of behavior that does not require too much preservation.

• Remove items that are very expensive or valuable.

• Identify a number of possessions that your person can freely play with.

• Fill in with comforting things that can make your person's life seem rich and full.

• Keeping a full compliment of playful possessions that can be lost and found can possibly prevent hoarding and taking things.

Hoards possessions and takes other's items

Hoarding is a common behavior in AD. It can range from never throwing out the trash, retrieving things from the trash, saving packaging from items, taking items from other places like day care and stashing items from all over the house. It does not matter who owns the item, a person with dementia will usually not consider that fact, just that the item may be needed now or in the future. My mother-in-law used to consistently take paintbrushes from the day care. Scolding did not work. I just quietly removed and returned them. We did let her keep some so she wouldn't take them all. Being

Notes:

non-confrontational and understanding can elicit more cooperation than being accusatory.

• Just observe the behavior, if you are so fortunate as to see it. Check the hiding place later and return the item to the right place when your person is not watching.

• Remember scolding will not change the behavior. It may just scare, annoy, or provoke a catastrophic emotional reaction. (The emotional centers of the brain are heavily affected by AD making them more prone to over reacting to negative situations.)

• If you observe your person taking a very valuable item and you don't want to lose track of it, make up an excuse to thank them for finding the object. Tell them how grateful you are and return it to its rightful place.

• If you are missing an item and did not see it taken, you may just have to wait until it turns up.

• You can try to observe your person's favorite hiding places and routinely check them for missing objects. Be careful that your person does not see you going through his/her possessions; this may set off a wave of paranoia or negative emotional reactions. If you are caught, don't act sheepish. Make up an excuse that you dropped something in the room. Even enlist your person's help in looking for your "lost" item. Be creative in these kinds of circumstances. You will make up your own tricks and ways of handling this problem.

Jewelry and fancy accessories

If your person is fascinated with jewelry, find some old pieces that you don't care about. It is likely that your person will pull the pieces apart but may have fun doing so.

• Remove items that are very expensive or valuable.

• Identify a number of pieces of jewelry that your person can freely play with.

• Fill up an old jewelry box just the way you would for a child's dress up collection. This will make your person's life seem rich and full.

• Compliment the collection with other dress-up items such as scarves, gloves, and purses.

• Visit the second hand store and purchase some of these items.

• Don't get upset or stressed if any of these items get hidden, lost or destroyed-that is what they are for.

• Keep just enough of these items so that you aren't tripping over them.

• Watch for unusual use of these favorite items—a dress as a towel-dipping brushes in the sink-flushing things down the toilet. You may have to limit some of the play items if you don't want your plumbing clogged.

Money

As with most things offer a substitute that gives comfort but is not a problem if lost or hidden.

• Try giving your person with AD fake credit cards.

• Try fake money and give them containers to store it in.

• Allow your person to count and freely play with the stash.

Notes:

• Give your person a large penny collection.

• Try one of those counting machines that sorts change. This may give your person countless hours of repetitious pleasure. Just like a young child who enjoys the security of repetition, your person may find this activity calming.

Clothing

This is much like jewelry and accessories.

• Remove items that are very expensive or valuable or that your fear will get ripped when your person tries to put them on backwards etc.

• Identify a number of pieces of clothing that your person can freely play with.

• Fill up an old basket or drawers with these items just the way you would for a child's dress-up collection. This will make your person's life seem rich and full.

• Compliment the collection with other dress-up items such as scarves, gloves, and purses.

• Visit the second hand store and purchase some of these items.

• Don't get upset or stressed if any of these items get hidden, lost or destroyed, since these clothes are dispensable.

• Keep just enough of these items so that you aren't tripping over them.

• Watch for unusual use of these favorite items—a dress as a towel-dipping brushes in the sink-flushing things down the toilet. You may have to limit some of the play items if you don't want your plumbing clogged.

Food

This is a trickier problem because you cannot fill up your person's room with food items. The source of this behavior may be very old feelings of food deprivation, or it may just be satisfying to have something left over for a snack in case of later hunger. This can represent a form of security. This behavior may occur in spite of all your clever tries to stop it. The worst thing that can happen is likely to be strange, rotten bits of things found tucked here and there.

• Offer meals with piles of small cut-up foods. This may make your person feel that this is a large and satisfying meal.

• Offer healthy snacks at reasonable intervals.

• Just let your person pocket the food and remove it later when they undress or have lost interest.

• Check your person's pockets, jackets, coats, and purses or bags for bits of stashed food.

• Don't argue or scold. Your person cannot learn not to do this behavior; it is merely a reaction to old feelings or immediate fears.

Other items

Treat this as you would any of the other items mentioned above.

• Remove items that are very expensive or valuable.

• Identify a reasonable amount of the favorite item that your person can freely play with.

• Fill up an old basket or drawers with these items just the way you

Notes:

would for a child's play collection. This will make your person's life seem rich and full.

- Compliment the collection with other like or companion items.
- Visit the second hand store and purchase some of these items.
- Don't get upset or stressed if any of these items get hidden, lost or destroyed, since that is what they are for.
- Keep just enough of these items so that you aren't tripping over them.
- Watch for unusual use of these favorite items—a dress as a towel—dipping brushes in the sink—flushing things down the toilet. You may have to limit some of the play items if you don't want your plumbing clogged.

Takes other people's possessions

This discussion has come into to play in other areas. Persons with Alzheimer's disease/dementia AD can develop a tendency to hoard and it is hard to determine which situations will stimulate the need to take things. If you know your person has this problem, be on the lookout for it.

- Prep the friends and relatives being visited about this problem. Reassure them that there is no need to be paranoid and that you will return any item that might be lifted.
- Always check your person's pockets or purse for unfamiliar items and return them to their owners.

PARANOIA

General

This emotional/mental health issue can be common in AD. It is hard to imagine how you might feel if there were no clear explanations for why things happen. You might also begin to believe people were after you and especially wanted to take your possessions, which keep disappearing on a regular basis. You could grow to blame others for most all of your difficulties. Scapegoats are needed and you and your family will most certainly get called upon to play that role. The key is to understand why this behavior happens and not to argue or chide. Your person is trying to make sense of his/her confused universe.

- Don't argue or confront your person with the "truth." It will only make matters worse and may provoke a catastrophic reaction.
- Encourage your person to see you as a confidant and to tell you his/her trusts and mistrusts.
- Be the trusted listener—not one of the troublemakers.
- Let your person vent his/her feelings.
- Offer empathy. Express understanding of how difficult it must be to have so many people that may want to "get" him/her.
- Give patient and calm explanations of expected events during an activity so that there is a minimum of surprises. This can prevent panic.
- Brush off any accusations aimed at you. Give a short explanation, if reasonable.
- Offer distractions to get your person's mind off of the problem.

Notes:

• Explain your person's tendency to paranoia to visitors in case they become targets. Reassure them that this is part of the disease and not to be taken personally.

• Decrease the amount of time spent in crowds as this tends to increase paranoia

• Use the suggestions in the: **Has catastrophic reactions** section (page 134); if your person gets extremely upset, let your person vent his/her feelings during a paranoid episode.

SOCIAL
RELATIONSHIP WITH FAMILY

General
In many ways the social aspect of health is often overlooked. Holistic health helps us focus on the complete person and not just the physical and mental aspects. These factors are admittedly overpowering in Alzheimer's disease/dementia (AD), but the disease affects all aspects of your person's life. The problem solving you will be called to do will be in all aspects of your person's life. The social aspects of life bring great joy and bonding along with misery and rejection. All of us can look back and see these factors playing out in our early life. As a person with AD travels back through time toward a childlike view of the world, we would hope that s/he can cast away some of the difficulties and maintain the social connections that give him/her happiness in his/her uncertain life. This can only be done with your help as the caregiver. Once you become the primary caregiver for your person, you are the gateway/conduit to the outside world. You will be making the decisions about visitors, social gatherings, as well as experiences like day care.

Able to recognize family members
This a great social time in your person's life s/he can still recognize people who have been important in his/her life. These relationships will drift away into feelings of familiarity without names and history. Now is the time to make connections and express love.

• Visit important relatives or encourage them to visit.

• Encourage small-scale gatherings that include important friends and/or relatives.

• Let your person go to parties/gatherings with you. It is amazing how your person can accommodate even if s/he is a bit forgetful. Your person may just not have the stamina s/he once had.

• Leave if your person becomes tired or visibly confused.

• Amend these circumstances to match your person's personality. If s/he is shy or reticent, then you will have to pick and choose social interactions that will be pleasurable but will not add too much stress.

• Take time to visit some adult day-care facilities. This may be a good time to begin thinking about day care. This can offer your person many social contacts plus activities and field trips that will be hard for you to create on your own. This will also give you respite time for errands, chores, or general down time.

Able to intermittently recognize family members
Overtime you will notice that your person will suddenly be unable to

Notes:

recognize people that s/he has known for most of his/her life. Just as quickly the recognition will return. The losses in this area of cognition fade in and out until one day the relative or friend will be someone else or simply unknown. Your person cannot control what is happening, these changes are due to damage to the brain caused by AD. The brain can no longer match what the eye sees with previous patterns created by memory (called Agnosia). In the beginning your person may realize that s/he doesn't know your person, but with time this understanding will be lost as well.

This loss can be one of the hardest for children, spouses, and relatives to bear. Without knowing it we can become very attached to being recognized. It is one of the great aspects of grief in dealing with your person. When this happens, allow yourself some private grief time to cry and feel the loss. It may take some time to make the needed personal adjustments. Be kind to yourself and your person. Caring, love and kindness are universal and are precious. Your person will still have feelings of familiarity with friends and family. Build on those responses and let that be enough.

• Reassure your person that forgetting people's names and faces is unfortunately part of the process and that you will help him/her through these awkward moments. This only applies if your person is in the early stages; otherwise try to gloss over the mistakes.

• Encourage all family members affected by this to freely grieve about it but not to blame your person. S/he cannot control what is happening. This problem is brain damage caused by the disease.

• Guide visitors out of the room if they have a negative reaction to what is happening. A negative reaction from a visitor could provoke an agitated or catastrophic response from your person. It is not worth it. The visiting person can return when he/she is composed and able to understand as well as cooperate with your explanations and requests.

• Once your person begins to routinely mistake the identity of relatives and friends, prep all people who come for a visit. Let them know what to expect and for the most part to play along with whatever name they are called.

• In the beginning relatives can try coming up to your person and (if appropriate) giving him/her a kiss on the cheek or holding your person's hands in theirs and saying something like "Hi Grandpa it's me Mary. How good you look. It's so great to see you." This can anchor your person and may work some of the time. If it confuses or agitates your person, then stop the practice.

• Try to act like things are normal when your person does not recognize a particular person.

• Try a matter-of-fact introduction.

• Don't argue if your person emphatically or rudely insists you are wrong. Just let it all go by and get the conversation on to something else.

• Remember that the social contact between familiar people matters more than the exact identity of each individual. Your person may quickly forget that a certain person even came to visit him/her. *The pleasure has to be in the moment.*

Responds well to a particular family member
This is an odd happening but a possibility. In the context of all things it

really does not matter if your person has a certain favorite. If that person is not you, then you may have to take some time to grieve and adjust.

• Invite this person over to visit your person on a routine basis. This could give your person pleasurable social interaction. You could even leave the two of them to play a game or just talk, and you can do chores or have respite time.

• Essentially work this to your advantage. This person can be a major source of respite for you.

• Be prepared for this attraction to fade. Many reactions to AD are individual and who knows where the damage will spread.

• Prepare the visiting person for the chance that one day s/he may be forgotten or arouse no positive reactions on a visit. Reassure your person that NOW matters and not to live in dread of a change that probably will happen.

Responds poorly to certain family members

Who knows why this might happen, it could be old memories cast on to a new person. If it does happen go with the flow.

• Gently find a reason to signal your person that is eliciting the response and escort him/her out. Reassure the visiting person that this is nothing to take personally and that this reaction may pass.

• Encourage your person to try again after a bit of time has past. Make a plan with your person as to how s/he will excuse him/herself if this reaction happens again.

• Try some distraction if a conversation goes awry. Suddenly offer some food/snacks to your person and visitor/s.

Often is rude and unthankful to family or friends

If this begins to be a consistent response, investigate and see if your person is having health problems. Is s/he in pain or discomfort? These questions may be hard for your person to answer, but make some attempt to get at the problem. You may need to take your person to the MD for a check-up to assist you in understanding why your person may be having this reaction.

If there are no medical causes, then your person may be experiencing some transitory confusion that has caused him/her to perceive an insult. This phase may well pass so try to flow with or around it.

• Don't engage the remarks or make angry comments in return. Continue on with whatever you are doing. The reaction or rudeness may just pass on its own.

• Don't argue or contradict. This is often hard to do but arguing will only aggravate the situation or cause a catastrophic reaction.

Routinely unable to recognize family members

Agnosia (inability to recognize people and objects) is getting more profound. Hopefully, by now you have gotten somewhat used to this situation and have worked with family and friends to make accommodations. If not then read the section: **Able to intermittently recognize family members** (page 148). Those ideas will still be usable. When this problem becomes routine you just need to step up your agility in responding to the strange names and titles you and others will be given. It is as if you are on the stage in a play and the cast of characters keeps changing. You may be asked to play a new

Notes:

role everyday. Not much is asked of you except to play along with the odd dialogue that can crop up. You may be asked how things are going at the "plant" where you both worked together.

• Go with the conversation and the role you have been cast in. Make up whatever you need to.

• Let your person guide the conversation and get answers to whatever questions that interest him/her.

• Follow the conversation carefully when it concerns your person's life history. You may find out some little known facts about your family. Not everything your person says is fiction, some may be truth told out of the current context. Your person can be "seeing" deceased relatives while talking to you as if you were an aunt or uncle. You can be involved in a little psychodrama and not even know it.

• Don't worry about truth. Your person is operating in his/her own your reality, not yours.

• Encourage visitors, whether family or friends, to follow along as well. Tell them they can follow your cue. You can guide the conversation and give them lead-ins as to what they can say.

• Treat the circumstance with a lighthearted view and encourage others to do so. Don't get hung up on the normal way things should work.

• Remember that the social contact between familiar people matters more than the exact identity of each individual. Your person may quickly forget that a certain person even came to visit them. The pleasure has to be in the moment.

Does not remember visits from family

This follows right along with the inability to recognize people and objects. Your person will probably not remember when s/he ate last and begin asking for more food. With all these issues, you have to remain calm and go with the flow. If your person was expecting a visit from family and the visit is over, s/he may still be asking when they are coming.

• Try gently reminding your person that the visit did occur and move on to describing how it went. See if this will distract your person onto the subject of what happened, etc.

• Try other methods of distraction and let the visit drop. The pleasure was in the moment.

• Don't let this stop you from arranging other visits. Your person will probably enjoy them regardless of not having memory of them.

Unresponsive to family members

At this point all that can be communicated is caring and love through calm words and simple body touch (i.e. face, hands or shoulders). Much of our communication takes place on this level, but we don't notice as much until the chattering of life goes away.

• Touch your person's shoulder and arm gently and then begin to speak, anchoring them in a connection with you.

• Explain who is visiting or what you are going to do (i.e. offering beverages or foods).

RELATIONSHIP WITH FRIENDS

General

Friends are an important part of most people's lives. You do not want to give up these connections for either of you, just because you are now a caregiver for a person with Alzheimer's disease/dementia (AD). Keep the door open on visitation. One-on-one visits or small groups are best for your person.

Able to recognize friends

This a great social time in your person's life, if s/he can still recognize people who have been important in his/her life. These relationships will eventually drift away into feelings of familiarity without names and history. Now is the time to make connections and express love.

- Visit important friends or encourage them to visit your person.

- Encourage small-scale gatherings that include important friends and/or relatives.

- Let your person go to parties/gatherings with you. It is amazing how your person can accommodate even if s/he is a bit forgetful. Your person may just not have the stamina s/he once had.

- Leave if your person becomes tired or visibly confused.

- Amend these circumstances to match your person's personality. If s/he is shy or reticent, then you will have to pick and choose social interactions that will be pleasurable but will not add too much stress.

- Take time to visit some adult day-care facilities. This may be a good time to begin thinking about day care. This can offer your person many social contacts plus activities and field trips that will be hard for you to create on your own. This will also give you respite time for errands, chores, or general down time.

Able to intermittently recognize friends

Overtime you will notice that your person will suddenly be unable to recognize people that s/he has known for most of his/her life. Just as quickly the recognition will return. The losses in this area of cognition fade in and out until one day the relative or friend will be someone else or simply unknown. Your person cannot control what is happening, these changes are due to damage to the brain caused by AD. The brain can no longer match what the eye sees with previous patterns created by memory (called Agnosia). In the beginning your person may realize that s/he doesn't know your person, but with time this understanding will be lost as well.

This loss can be very hard on friends who hope that they are bringing personal cheer with their visits. Dealing with the need to be recognized can be very difficult. Preparation is important. Take some time to encourage friends to grieve their losses and to make the needed personal adjustments so that they can continue to visit. Remind them that the pleasure is in the moment and they are making a gift to your person even if s/he cannot remember. Caring, love and kindness are universal and are precious. Your person will still have feelings of familiarity with friends and family. Build on those responses and let that be enough.

- Reassure your person that forgetting people's names and faces is unfortunately part of the process and that you will help him/her

Notes:

through these awkward moments. This only applies if your person is in the early stage; otherwise try to gloss over the mistakes.

• Encourage all friends affected by this to freely grieve about it but not to blame your person. S/he cannot control what is happening. The disease and the damage to the brain cause this problem.

• Guide visitors out of the room if they have a negative reaction to what is happening. A negative reaction from a visitor could provoke an agitated or catastrophic response from your person. It is not worth it. The person can return when he/she is composed and able to understand as well as cooperate with your explanations and requests.

• Once your person begins to routinely mistake the identity of relatives and friends, prep all people who come for a visit. Let them know what to expect and for the most part to play along with whatever name they are called.

• In the beginning friends can try coming up to your person and (if appropriate) giving him/her a kiss on the cheek or holding your person's hands in theirs and saying something like "Hi Anne it's me Mary. How good you look. It's so great to see you." This can anchor your person and may work some of the time. If it confuses or agitates your person, then stop the practice.

• Try to act like things are normal when your person does not recognize a particular person.

• Try making a matter-of-fact introduction.

• Don't argue if your person emphatically or rudely insists you are wrong. Just let it all go by and get the conversation on to something else.

• Remember that the social contact between familiar people matters more than the exact identity of each individual. Your person may quickly forget that a certain person even came to visit them. *The pleasure has to be in the moment.*

Responds best to a particular friend

This is an odd happening but a possibility. In the context of all things it really does not matter if your person has a certain favorite.

• Invite this person over to visit your person on a routine basis. This could give your person pleasurable social interaction. You could even leave the two of them to play a game or just talk, and you can do chores or have respite time.

• Essentially work this to your advantage. This person can be a major source of respite for you.

• Be prepared for this attraction to fade. Many reactions to AD are individual and who knows where the damage will spread.

• Prepare your person involved for the chance that one day s/he may be forgotten or arouse no positive reactions on a visit. Reassure your person that NOW matters and not to live in dread of a change that probably will happen.

Has friends that live out of area

This is a bit more difficulty because the exchange cannot take place in person. If these individuals express concern, encourage them to stay in touch. Their connection is also a gift and can bring pleasure to your person whose external world is slowly shrinking.

- Encourage these friends to write letters or e-mails as a way of keeping in touch.

- Encourage these friends to write friendly chatter about their lives. Although your person may no longer remember who this person is s/he may enjoy the stories in the letters.

- Read these letters and e-mails to your person as part of an evening entertainment. This activity can bring moments of pleasure and connectedness.

- Play up the part that people care about your person, and through that let your person feel loved and cared about.

- Embellish as needed to keep your person interested.

Unhappy that friends never visit

Reassure your person that s/he is loved and that friends have busy and difficult lives that make it hard for them to be available.

- Encourage brief visits whenever feasible.

- Encourage friends to write or e-mail and follow the steps outlined above.

Responds poorly to certain friends

Who knows why this might happen; it could be old memories cast on to a new person. If it does happen go with the flow.

- Gently find a reason to signal your person that is eliciting the response and escort him/her out. Reassure your person that this is nothing to take personally and that this reaction may pass.

- Encourage your person to try again after a bit of time has passed. Make a plan with your person as to how s/he will excuse him/herself if this reaction happens again.

- Try some distraction if a conversation goes awry. Suddenly offer some food/snacks to your person and visitor/s.

Often is rude and unthankful to friends

If this begins to be a consistent response, investigate and see if your person is having health problems. Is s/he in pain or discomfort? These questions may be hard for your person to answer, but make some attempt to get at the problem. You may need to take your person to the MD for a check-up to assist you in understanding why your person may be having this reaction.

If there are no medical causes, then your person may be experiencing some transitory confusion that has caused him/her to perceive an insult. This phase may well pass, so try to flow with or around it.

- Encourage friends not to engage the remarks or make angry comments in return. Continue on with the conversation. The reaction or rudeness may just pass on it's own.

- Encourage friends not to argue or contradict. This is often hard to do but arguing will only aggravate the situation or cause a catastrophic reaction.

Routinely unable to recognize friends

Agnosia (inability to recognize people and objects) is getting more profound. Hopefully, by now, you have gotten somewhat used to this situation and have worked with family and friends to make accommodations. If not then read the section : **Able to intermittently recognize friends** (page 151). Those ideas will still be usable. When this problem becomes routine you

Notes:

just need to step up your agility in responding to the strange names and titles you and others will be given. It is as if you are on the stage in a play and the cast of characters keeps changing. You may be asked to play a new role everyday. Once you accept that these parts are just reflections of one character- you. Not much is asked of you except to play along with the odd dialogue that can crop up. You may be asked how things are going at the "plant" where you both worked together. Fill friends in on this surreality when they come to visit.

- Prep friends to go with the conversation and the role they have been cast in. They will need to make up whatever is needed.

- Let your person guide the conversation and get answers to whatever questions that interests him/her. Prep the friends to this actuality.

- Prep the friends no to worry about truth. Your person is operating in his/her own your reality, not theirs.

- Encourage visitors, whether family or friends, to follow along as well. Tell them they can follow your cue. You can guide the conversation and give them lead-ins as to what they can say.

- Treat the circumstance as fun and encourage others to do so. Don't get hung up on the normal way things should work.

- Remember that the social contact between familiar people matters more than the exact identity of each individual. Your person may quickly forget that a certain person even came to visit them. The pleasure has to be in the moment.

Unresponsive to friends

At this point, all that can be communicated is caring and love through calm words and simple body touch (i.e. face, hands or shoulders). Much of our communication takes place on this level, but we don't notice as much until the chattering of life goes away.

- Touch your person's shoulder and arm gently and then begin to speak, anchoring them in a connection with you.

- Explain who is visiting or what you are going to do (i.e. offering beverages or foods).

OTHER SOCIAL RELATIONSHIPS

General

All types of social relationships are important. Keep your connections up and don't run the risks of being socially isolated. It will make caregiving more enjoyable when you have friends to support the both of you.

Able to be social in groups

If this is possible, encourage some group activities.

- Encourage small-scale gatherings that include important friends and/or relatives.

- Let your person go to parties/gatherings with you. It is amazing how your person can accommodate even if s/he is a bit forgetful. Your person may just not have the stamina s/he once had.

- Leave if your person becomes tired or visibly confused.

- Amend these circumstances to match your person's personality. If s/he is shy or reticent, then you will have to pick and choose social interactions that will be pleasurable but will not add too much stress.

• Take time to visit some adult day-care facilities. This may be a good time to begin thinking about day care. This can offer your person many social contacts plus activities and field trips that will be hard for you to create on your own. This will also give you respite time for errands, chores, or general down time.

Has difficulty being social in groups

This is natural for certain personalities and at certain stages of the disease. Not much accommodation needs to be made unless you are a highly social person who likes to have parties in your house.

• Let your person stay in his/her room with favorite activities or TV to entertain.

• Check on him/her periodically to see that all is well.

• Hire a babysitter to keep up with your person's needs and entertainment. You will be in the vicinity so questions can be answered. You may not need a professional for this.

Will not participate in group activities

This may not be a permanent situation. Your person may be shy and have a hard time making friends. You can still consider day care. Even people who seem to like lone activity can find pleasure in working alongside others. Group energy can bring comfort. If your person is fearful in a group, you may have to keep them in your home and create activities for him/her to do. Remember that young children learn to play and work in a group at a young age (i.e. kindergarten). Your person may relive some of these experiences and find comfort in them.

• Don't give up on group activities like day care. Encourage your person to stay and try it for a while. S/he may grow to like it even if s/he never becomes a highly social member.

• Day care workers are trained and encouraged to bring reticent members into the group and to keep him/her from feeling left out.

Makes inappropriate social advances

As various parts of the brain are affected by dementia, unusual things can happen. You are not usually privy to all the experiences your person has had in life. If your person has been a recipient of inappropriate advances or sexual abuse in his/her life, s/he may "act out" at some point in the progress of the disease. Or s/he can be tapping into natural curiosity that young children have. Where they may be inclined to touch or hug people impulsively before invading another person's personal space and that these actions have consequences. Your person is likely at this point to be unable to grasp the consequences of his/her actions unless there is an immediate yell or pushing way by your person who is the object of the advance. Either way, unless there is strong evidence to the contrary, you can assume that your person is probably acting out of naiveté or confusion and not out of a desire to hurt anyone else.

• End the circumstance immediately but gently and firmly. This way your person will not go into a confused agitation or catastrophic reaction.

• Research the circumstance looking for triggers to the behavior.

Notes:

Notes:

• Consult your physician if you suspect your person has been subjected to child abuse or sexual abuse.

Unresponsive in social situations

At this point you will most likely not be encouraging your person's participation in social events.

If you still feel the need to keep up your social connections through parties, then make sure your person's needs are taken care of and carry on.

• Hire a babysitter to keep up with your person's needs and entertainment. You will be in the vicinity so questions can be answered. You may not need a professional for this.

SOCIAL BEHAVIORS

General

Your person's reactions to other people will vary over time and through the progress of the disease. You have to be prepared for changeability. Remember not to take your person's negativity or rudeness personally. These comments are just a way for your person to cope and impose some level of control on his/her confused life.

Often is pleasant and polite to acquaintances

This is a great social time for your person. S/he may be able to converse at parties just through pleasantries and nodding the head. It is amazing how far this type of communication will carry you.

• Encourage parties and group connections.

• Follow ideas in the sections on **RELATIONSHIPS WITH FRIENDS** (page 151)

Often is snappy, rude, or hostile to acquaintances

Time is going by and your person may have had a personality turn for the worse. When dementia affects the limbic system or the emotional center of the brain the results can be labile or changeable emotions and emotional reactions. This problem may pass or you may have to learn to work with it.

• Encourage friends or acquaintances not to engage the remarks or make angry comments in return. Continue on with whatever the conversation is. The reaction or rudeness may just pass on it's own.

• Encourage friends not to argue or contradict. This is often hard to do but arguing will only aggravate the situation or cause a catastrophic reaction.

VISITING

General

Visiting is a challenging part of maintaining social connections. Your person's reactions to other people in visiting situations will vary over time and through the progress of the disease. You have to be prepared for changeability. Remember not to take your person's negativity or rudeness <u>personally</u>. These comments are just a way for your person to cope and impose some level of control on his/her confused life. Try to keep these social connections in place for as long as you can.

Can go for visits to relative or friends house

If your person is pleasant and portable, take him/her for periodic visits to the houses of friends and relatives.

• Encourage small-scale gatherings that include important friends and/or relatives.

• Let your person go to parties/gatherings with you. It is amazing how your person can accommodate even if s/he is a bit forgetful. Your person may just not have the stamina s/he once had.

• Leave if your person becomes tired or visibly confused.

• Amend these circumstances to match your person's personality. If s/he is shy or reticent, then you will have to pick and choose social interactions that will be pleasurable but will not add too much stress.

Often tries to take other people's possessions

This discussion has come into to play in other areas. Persons with Alzheimer's disease/dementia (AD) can develop a tendency to hoard and it is hard to determine which situations will stimulate the need to take things. If you know your person has this problem, be on the lookout for it.

• Prep the friends and relatives being visited about this problem. Reassure them that there is no need to be paranoid and that you will return any little item that might be taken.

• Always check your person's pockets or purse for unfamiliar items and return them to the owners.

• Don't chide or embarrass your person if you do find items. Understand that s/he cannot control this behavior and just make accommodations.

• Smooth over the occasion if you catch your person in the act of starting to pocket something that doesn't belong to him/her. Gently take the item from your person and openly admire its beauty. Compliment your person and host for his/her good taste. Return the item to its proper place.

• You can quietly apologize later if you feel the need. If you have prepped the friend or relative, this will probably blow over and give you both a good laugh.

RESPITE

General

Respite is a key supportive factor for your survival as a home caregiver. Even if you try to do everything for your person, you will need to leave him/her at some point. This is a similar transition that parents make when they have to begin using babysitting for their small children. It is nerve wracking at first worrying about how the kids are doing while out of your sight. Relax—most adults and children make it through these transitions without harm. It is a good idea to get used to using respite before you have to use it in critical situation. It saves you added worry.

Can go for overnights with friends or relatives

This is wonderful for you as a primary caregiver. If you have immediate relatives or friends who can help you at this level, then you have a powerful

Notes:

---

support network. Read **CHAPTER 2** on building and utilizing your support network for ideas on how to maximize the use of potential offers of help.

- Prep all hosts as to the general needs of your person, especially food preferences.

- Prep all hosts as to any idiosyncrasies or behavioral difficulties your person may have.

- Leave contact phone numbers as you would for any sitting situation.

- Pack a reasonably small bag of clothing and comfort items.

- Hand off this book or another copy to your prospective hosts. Mark where your person is in all the domains. Inform your hosts on how to use this book and its use in getting ideas on handling any odd situations that might come up.

- Write any special reminder for your host sitter in the back of this book and point it out to them.

- Keep new notes after each situation, so that future problems can be avoided and your person can have a more enjoyable return visit.

Can handle extended stays with friends or relatives

This situation is not too much different than setting up for an overnight. Time can be tricky for your person. If s/he is with familiar relatives or friends, the whole time may go very smoothly. This will depend on where your person is in his/her disease process. Sometimes you will not know if it will be a problem until you try an excursion like this.

- Try a short stay first, if possible before going for an extended stay. If the host lives far away, this may not be possible; and you will have to give it your best prep and pray. Transportation may be a bigger issue.

- Read the section on **VACATIONING** (page 162) before making plans to ship your person off to a distant destination for a prolonged visit. Otherwise you may learn the hard way (I did).

- Follow all the suggestions in the section above on the overnight stay.

- If this friend or relative is going to be offering respite on a frequent basis, consider getting them a copy of this book and copy all the information you have marked in yours. This way you can trade suggestions and notes. You will both be speaking the same language.

Becomes agitated or restless or periodically becomes a behavior problem

This makes it difficult for both you and your possible respite providers. At this point keep the visits down to several hours. If the host has some experience at this, then overnights may still be a possibility. Longer stays will test the fortitude of your host and are probably not a good idea unless you have been working with this host for a while and s/he is skilled in caring for your person.

- Prep all hosts as to the general needs of your person, especially food preferences.

- Prep all hosts as to any idiosyncrasies or behavioral difficulties your person may have.

- Leave contact phone numbers as you would for any sitting situation.

- Hand off this workbook or another copy to your prospective hosts. Mark where your person is in all the domains. Inform your hosts on

158   *Alzheimer's Workbook*

Notes:

how to use this book and its use in getting ideas on handling any odd situations that might come up.

• Write any special reminder or instructions for your host sitter in the back of this book and point it out to them.

• Keep new notes after each situation, so that future problems can be avoided and your person can have a more enjoyable return visit.

• Have your hosts call if they get into a difficult situation. If you all cannot brainstorm a good solution, then return and pick up your person. You don't want to create ill will or burn your bridges.

RECREATIONAL ACTIVITIES

General

Much pleasure can be derived from enjoyment of recreational activities. This will facilitate your person's enjoyment of group events and day care activities. The main idea is to keep your person's brain stimulated at a pleasurable or tolerable level to preserve general function and make care easier.

Many people aren't even sure if they like these activities because work has defined most of their lives. I have seen a day care offer envelope stuffing and licking as a recreational activity to suit this very need. If your person is in the early stages of Alzheimer's disease/dementia (AD) and is not sure about recreational activities, try some out. Activities can range from painting to crafts and puzzles or music appreciation. All of these can be adapted over time to match your person's abilities.

Enjoys recreational activities

Recreation is essentially having fun and enjoying life. It may take a bit of work to match your person's skills with activities that will allow him/her to have the most enjoyment.

• Help your person keep up with any definite arts or crafts that s/he has already learned and has maintained as a hobby or an art. The famous painter, De Kooning continued to paint for a number of years after he was diagnosed with AD. Adaptations can be made.

• Rekindle activities that your person may have done in the past but are not currently involved in. If your person needs help refreshing a skill, think about inviting a local artisan to come in and offer some one-on-one coaching. This person can also give you ideas on what adaptations may be needed now or in the future. This is especially helpful for fabric arts like knitting.

• Identify your person's favorite activities while s/he can still converse about them and can give you feedback.

• Try various types, if s/he is unsure what might be a favorite.

• Go for low to moderate complexity. Keep things as simply as possible. You may not be sure what skills your person may still have.

• Work with feedback from your person. If s/he becomes frustrated at a certain level then simplify.

• Let your person get started at a day care facility and let the staff help you understand your person's likes and dislikes and what might be adapted for home recreation.

Notes:

• Don't forget singing, dancing, and listening to or playing music. Dancing is often done in day cares or senior centers.

• Be gentle and tactful with your person as his/her cognitive abilities fade. The decline in motor skills will eventually make many hobbies difficult. Encourage simple participation and overlook flaws. Praise the results.

• Toys can be acceptable as recreation. Some people worry that they make a person with AD seem too childlike, but inside your person can be seeing the world through a child's eyes. Stuffed animals can be a favorite. Use trial and error. Give them to your person as gifts and see if s/he pays any attention.

• Remember that recreation may take a bit of energy to set up, but it can provide you with respite when your person is happily occupied and you can get a few chores done.

• Read the section on: **GARDENING** (page 101) if that is something your person is or might be interested in.

• **Refer to the list under favorite recreational activities (on the next page) for some ideas.**

Enjoys work activities as recreation

This takes a bit of creativity but you should not be put off by it. Think about how you may have set up activities for children. Schools set up kitchens and work desks for kids to mimic adult actions. Your person is doing things in reverse. You can create special environments for your person.

• Use the kitchen as a setting if your person was a housewife or cook. You will have to supervise, but there is plenty of opportunity because meals have to be cooked everyday. Refer to the section on **COOKING** (page 84) for ideas and safety issues.

• Set up a desk in your house. Stock it with items that are familiar to your person. Let him/her arrange and rearrange these items. Provide paper, pencils, envelopes, fake stamps and markers (watch this one because just like children, persons with dementia can mark on walls and furniture at a certain stage in the disease). You can evolve this into art and reading area if some particular activity catches on.

• Set up a safe carpentry area if your person was an engineer, plumber, electrician, carpenter, or fix-it kind of person. This will take some work to make sure there are few safety hazards. If this is not your skill, call on a friend or relative to come in and help you set this up. Don't worry that this may be imposing. Often people are looking for a way to help you with the skills that they have. Read **CHAPTER 12** on **Accepting and building support**. This person may even be willing to come for routine visits and work with your person on various projects. More respite time for you.

• Read the sections on **WORKING AROUND THE HOUSE OR SHOP** section (page 97) to develop more ideas for recreational work activities.

Favorite recreational activities: Mark the ones your person enjoys and make notes.

Artwork (calligraphy, painting, drawing, cutting, tracing, and pasting)

Bird watching

Caring for pets or plants

Notes:

Cooking

Dancing

Gardening

Housework

Knitting

Making scrapbooks

Pottery

Reading books and magazine

Sewing

Stamping

Stamp collecting

Likes to play with certain toys:

Listening to music

Listening to the radio

Listening to books-on-tape

Looking at picture albums or books

Playing musical instruments

Playing puzzles and games

Watching TV (may think people on TV are in the room)

Working with photo or memoir albums

Work-related activities

Writing poetry, memoirs, thoughts

Does not enjoy recreational activities

This is a more difficult situation because boredom could become a problem. It may grow into a behavioral issue if your person does not have things to occupy his/her time. If your person does not like arts or crafts, look for other non-traditional ideas.

• Identify any activity your person would consider pleasurable. Talk with your person about why certain activities make him/her happy or not. You may be able to find something that can be fashioned into a recreational activity.

• Look into areas of work that your person did for a number of years. Can any of these be adapted into a recreation? A scientist who didn't like much of anything, found happiness in working with some home made picture puzzles.

• Look for household activities that can be done repeatedly and possibly give comfort or pleasure (i.e. folding laundry, pushing a sweeper, carpentry).

• Don't forget singing, dancing, and listening to music.

Refuses to participate in recreational activities

If your person is refusing to do recreational activities, s/he may be depressed or moving into a more passive state. If you are worried about depression see your MD for an evaluation. If this decline in interest seems like a natural state, then go with the flow and adapt. At this point TV, radio, music, or mock reading may be all your person will do. In the earlier stages encouraging

Notes:

brain function through stimulation is important. As your person becomes more passive and moves toward the end stage of the disease, stimulation becomes less important than comfort and security.

VACATION

General

This is a time for your person to get in some vacation enjoyment as the opportunities will decrease in the future as his/her cognitive abilities decline. Your person will still do best if chaperoned. Mild transient memory loss will not be too problematic if your person is with some one who can fill in the blanks, but it could cause quite a bit of worry if he/she is traveling alone. The unpredictability of the memory loss could create the sudden problems with increased confusion, anxiety and fear.

Can handle vacation trips

• Choose easy tours or vacations with plenty of support staff. Do not expect other tour guests to help you with your person. You may need to pay extra for disability supports.

• Plan easy daily itineraries and keep a slow pace. This will minimize his/her confusion from being overly tired.

• Try to minimize external stimuli. A day or so here and there is ok. Too many stimuli (i.e. noise, crowds, and visuals) can promote confusion and exhaustion. For instance a place like Disney World could go both ways, on a quiet day without large crowds the busy colors and fast moving characters could be tolerated. If the place is crowded and noisy, the stimulation could be overwhelming. You can judge by your person's reaction what s/he is able to tolerate.

• Try cabins or getaways where you stay in one place for a number of days. These can be low stimuli and relaxing. Nature is a plus because of the healing connections with the numinous feelings of things universal or greater than human.

• Accompany your person on walks in unfamiliar places. S/he may be capable of finding his/her way back, but it may be too great a chance to take.

• Keep an ID card on your person whenever you travel just in case you get separated. Enroll in the Alzheimer's Association's Safe Return program as a back up for US travel.

• Do all the packing yourself. This will minimize confusion and forgotten items.

• Have your person dress in comfortable clothes for easier toileting.

• Pack up all the normal eating supplies that you would be using at home. Pack a snack pack that can include an adult bib and special eating utensils if you need them (a spill proof cup or cup with built in straw could be very handy for offering fluids in a car) as well as bite-size food.

• Pack any favorite comfort or activity items that will help calm your person or make the time pass faster if you are taking a long car journey. Music or stories on cassette or CD can help to pass the time.

• If you are traveling by car, be sure you have emergency equipment and/or have enrolled in a roadside emergency service.

• Never leave your person alone in a car. S/he could become confused and release the brake, etc.

• Persons with dementia should not drive, so do not plan to have them spell you.

• If you are staying in a hotel, be alert to wandering. You can obtain door alarms on-line from: www.alzstore.com. You can also travel with a doorknob safety cover, which will keep your person from getting out the door (be sure and check to see what kinds of knobs these work on). Also, you sleep in the bed closest to the door.

• Check for safety hazards wherever you go. Unplug coffee pots and hairdryers that may cause problems for a curious person with dementia.

• Have flexible contingency plans so that you can cut your vacation short if your person is not tolerating the trip well or becomes ill.

• Good planning and full consideration of safety issues will help to ensure a pleasant vacation.

• Plan rest periods into any daily vacation schedule.

• Keep your sense of humor. It will greatly enhance your vacation.

May be able to travel alone

Even if your person seems to have most all of his/her faculties, important losses have occurred. By the time the Alzheimer's disease/dementia (AD) diagnosis has been given, memory loss will be intermittent. If your person gets into a difficult situation and has a memory loss, the resultant confusion could make the situation critically dangerous. You and your family will have to decide how big the risk is and how to minimize it. <u>In general, try not to let your person with AD travel alone.</u>

• Be careful with lone travel on planes. The airlines do not usually allow the elderly to travel in the care of a stewardess, but they may be willing to offer a bit of extra support. You can also ask for a wheelchair transport between planes if you cannot arrange a direct flight. Watch out for transitory confusion. A person with AD can be very convincing when claiming that his/her ticket is wrong and s/he should not go to that destination (my mother-law-did this and caused quite a turmoil).

• Try to arrange direct flights if you have to risk a lone travel. New airport regulations no longer allow you to meet someone at the gate. So have your person met with a wheelchair and delivered to you or your family who are waiting at the entrance to the gates. You do not want your person to get confused and start wandering in the airport or get picked up by security personnel (this would be preferable to having your person get harmed).

• Keep an ID card on your person whenever you travel just in case s/he gets lost. Enroll in the Alzheimer's Association's Safe Return program as a back up for US travel.

Unable to fly alone but still able to fly

This will mean that you, a relative, or a friend will have to fly with your person. This is actually the safest way for your person to travel because of the unpredictability of memory loss and behavioral adaptations.

• Arrange for a direct flight if possible and plan to fly at a time when your person is at his/her best.

Notes:

- Avoid traveling when the airports will be the most crowded (i.e. holidays) and when there is a good chance of layovers due to bad weather.

- Explain the plan to your person as you pack up and get ready to leave. Your person may well forget what you said but in some way a shred of it may stick. You can build on that, as the time gets closer. Do not be surprised if s/he asks again and again where you are going and what you are doing. Be patient and give short repeated explanations.

- Use distraction if your person gets fixated.

- While traveling give short explanations of each thing that is going to happen right before it happens. (i.e. "we are going to get on the plane now or we are going to walk this way and find our seats"). This is essentially what you would do if you were traveling with a small child. These calm explanations will decrease possible confusion.

- Bring a favorite activity for your person to pass the time more easily. Check with the airlines before bringing knitting needles on board or any tools that might violate regulation.

- Pack recreational items into a kit so that they can be inspected more easily.

- Pack a snack pack that can include an adult bib and special eating utensils if you need them (a spill proof cup or cup with built in straw could be very handy for offering fluids during a bumpy ride) as well as bite-size food. This will be helpful if your person gets hungry in between airplane meals. Your food may be more appropriate than what is being offered anyway.

- Be aware of fluid needs verses bathroom needs. You can control the amount of water your person consumes and use juices, which may be processed more slowly.

- Do not serve your person alcohol on the flight. Even if s/he may handle some on the ground, you may be asking for trouble in the confined circumstances of an airplane.

- Pack some handi-wipes for cleaning up you and your person after any meals. This will spare you from having to ask for towels from the stewardess.

- Pack all this extra stuff in a small roll-on bag if it won't fit in a shoulder bag. Try to position the bag under the seat where you can reach everything.

- Keep all important papers in your bag (i.e. passports and tickets.) Your person can simply carry a duplicate ID.

- Using wheelchair services may be helpful to get quickly between gates even if your person can walk.

- If you have a layover, see if the airport has lounges for disabled persons (your travel agent may be able to arrange access for you.)

- Pre-board with the small children and others with disabilities.

- If possible ask to be seated with an extra seat next to you. It is great if you have the seat for all your supplies.

- Seat your person in the window seat so that s/he can look out the window and interact only with you.

• Whenever possible have someone meet your plane. You will be happy to have the extra hands after even a short flight.

Unable to go on vacation with you and/or your family

You will need to identify friends or relatives that your person is familiar with and ask them if they can handle an extended visit. Hopefully you will find someone who has been anxious to make a contribution. It would be ideal if they have cared for your person before so the extended stay will not pose many new behavioral issues. It is also great if it can be a sister or brother with whom your person is deeply familiar. Often this familiarity is something your person can fall back on if s/he forgets the name of your person with whom s/he is staying. In the end the success of the whole matter will depend on where your person is in his/her disease process. Sometimes you will not know if it will be a problem until you try an excursion like this.

• Try a short stay first, if possible before you go for an extended stay. If the host lives far away this may not be possible and you will have to give it your best prep and pray. Transportation may be bigger issue.

• Make sure your person has safe transportation to and from his/her destination.

• Consider getting the host family a copy of this book and copy all the information you have marked in yours. This will give them a head start in problem solving behavioral issues that come up while you are gone.

Unresponsive to vacation settings

At this point it would be torture for you and your person to go on vacation. The time for vacation has passed. Your person is now more concerned about comfort, good care, and security. If your must leave, see the above section on having to leave your person behind.

• Try all the previous ideas about extended respite.

• Consider Nursing or Rest home respite if your person's care is extensive and s/he is in the late stage of AD. This care can be expensive, but well worth it if you just have to get away.

SPIRITUAL
BELIEFS

General

This is the fifth dimension of the whole person and one that is considered the least. We spend much time worrying about our physical, mental, emotional and social health, but do not consider that our spirit is an inseparable part of the equation. The orientation of this section is to a belief in God or a Supreme Being but loyalty to no particular religion. The chapter on Alzheimer's disease and Spirituality contains a more extensive philosophical discussion and only the brief essences will be mentioned in this section because it is about the more practical aspects of incorporating spirituality into daily living.

Believes in God, Supreme Being, or Universal Divine State

Even if your person is not very talkative about spiritual beliefs or matters, encourage a discussion from time to time. We do not know the true interface between this disease and the growth of the soul, but there have been

Notes:

speculations that there may be an individual spiritual path to follow in which the mortal soul is seeking to unite with the immortal soul and the energy of God. The religion/spiritual beliefs a person has followed in his/her lifetime contain stepping stones on this path. No one can predict the twists and turns, philosophical changes or spiritual insights that your person may experience. As a caregiver keep your mind open and proceed by intuition. There are no true rights or wrongs if you are guided by love and work with your person to strengthen his/her spiritual connections. You can be the channel and facilitator to the divine. Any activities done in this manner will strengthen your own spiritual connections. You do not know the blessings that may come into your life for having given this gift to your person. This is your chance to express ultimate compassion from one soul to another.

• Do not force a direction or preach heavily to your person. S/he cannot learn but only make feeling connections to spiritual material.

• Remember that in this area it is not your agenda that matters, but what you can offer your person to facilitate his/her spiritual connection.

• Set aside time every week, if you can, for a short spiritual session with your person. If you have something you already do for yourself or a religious activity you could, if appropriate, involve your person in, use that circumstance as a starting point. If you feel the need to preserve your time for yourself make another time for your person.

• Follow the rituals of your particular religion if you have one. This may be sufficient for you and your person, but consider setting up special spiritual sessions for your person as well. When s/he can no longer attend religious ritual, these sessions could be of comfort.

Setting up spiritual sessions

• Make a library of spiritual books and tapes that may be meaningful to you or your person. Collect them as you have a chance. Try all kinds of different types if you are so moved. Christian and Spiritual bookstores are wonderful sources of material. (See the appendix on Spiritual Material for a list of some things your might try).

• Collect or rent spiritual video tapes as well. Your person does not have to connect with the exact words or visuals in the material to make the spiritual connection.

• Try new things interspersed with old favorites. If you are not entirely sure what your person will like, pick things that will please you. Your positive feeling state will energetically be transferred to your person and may bring happiness to your person as well.

• Use the feedback from your person about his/her favorites and set those aside in a special section.

• Extend compassion to yourself if you stumble across something your person does not enjoy. It does not matter. There are no rights or wrongs, make notes and move on.

• Bibles can be tagged with favorite passages if this is what your person enjoys.

• Don't overlook poetry, operatic or classical music, and time in nature as spiritual tools. It is all in your attitude and presentation that can bring out the sacred.

• Pre-select some materials for your session, giving yourself several

Notes:

choices because it will be hard to be intellectual and make choices from large amounts of material when in a feeling/spiritual state.

• Access a feeling state by the method described below and proceed with your session. This feeling state is an openhearted state that can enhance your connection with the divine. If it is too hard or disturbing in some way to go into a feeling state, then start you session at your own comfortable place.

Accessing and working in a spiritual state

• Think about love and those people you love and who love you. Surround yourself with these feelings and ground yourself in your own spiritual connection before making exact decisions about the spiritual materials that you will use in your session. This will stimulate your intuition and what you may pick is very likely to be what is needed. You will know you are in this spiritual state because you will feel the center around your heart open. You may even be moved to tears as you proceed into this powerful feeling state. This is a holy connection you are forging.

• Open your session with some spiritual music. This could be your whole session or a method to set the tone and help you access the feeling state of love.

• Choose the prayers, readings, or music you may want to use during your spiritual session. Use the compassionate heart state to feel the material before you engage it. This may guide you to a certain passage.

• Encourage your person to express thoughts and feelings about the materials. Do not be afraid to engage large or small questions or issues your person may wish to discuss. You don't have to have any answers. If your person gets lost in the question, just move on. These sessions are more about connecting and feeling than intellectual discussions.

• Do not be surprised if your person has moments of spiritual lucidity or understanding. The inspiration for this comes from a place other than normal cognition and is connected with the divine. If your person does open up, consider making notes after the session or tape recording sessions (if your person does not mind-don't let the taping intrude).

• Consider ending each session with more spiritual music for calming closure. During this time you can rest yourself and let the energy of the session flow through you and restore you.

• Keep a journal of your activities and what was discussed. This could be a wonderful record for you and other generations after your person has passed on. Make special notes of spiritual requests around end-of-life issues.

• Create a Spiritual Signature for your person by making a list over time of his/her favorite music, pictures, colors, and spiritual passages that may be used to sooth your person now and at the end of his/her life. At that time you will have the chance to create the ultimate spiritual session as your person's soul passes on to the realm of God.

• Pray routinely for your person (if it is in your belief system). This loving energy and intention can help to bring peace to you and your person.

• Read the section on death. It is an important subject that should be included in your sessions.

Notes:

Does not believe in God, Supreme Beings, or Universal Divine State

If you or your person do not believe in any divinity or have great doubts about an afterlife and would never consider reading spiritual texts, do not dismiss this section without thinking of adapting this idea into a concept of earthly pleasure. No harm will be done. Consider setting aside time to have art or cultural sessions. These could be soothing and relaxing for your person and for you.

• Set aside time every week, if you can, for a short cultural session with your person. If you have something you already do for yourself you could, if appropriate, involve your person. If you feel the need to preserve your time for yourself, make another time for your person.

• Make a library of great books, recreational reading and tapes that may be meaningful to you or your person. Collect them as you have a chance. Try all kinds of different types if you are so moved. Regular and Spiritual bookstores are wonderful sources of material. (See the section: **SELECTED READINGS** (page 225) on **Spirituality and Inspiration** for a list of some things you might try)

• Collect or rent video tapes as well. Your person does not have to connect with the exact words or visuals in the material to make an enjoyable connection.

• Try new things interspersed with old favorites. If you are not entirely sure what your person will like, pick things that will please you. Your positive feeling state will energetically be transferred to your person and may bring happiness to your person as well.

• Use the feedback from your person about his/her favorites and set those aside in a special section.

• Extend compassion to yourself if you stumble across something your person does not enjoy. It does not matter. There are no rights or wrongs, make notes and move on.

• Books can be tagged with favorite passages if this is what your person enjoys.

• Consider poetry, operatic or classical music, and time in nature as cultural tools.

• Open your session with some special music. This could be your whole session or a method to set the tone.

• Encourage your person to express thoughts and feelings about the materials. Do not be afraid to engage large or small questions or issues your person may wish to discuss. You don't have to have any answers. If your person gets lost in the question, just move on. These sessions are more about connecting, feeling, and enjoyment rather than intellectual discussion.

• Do not be surprised if your person has moments of lucidity or understanding. The inspiration for this comes from a place other than normal cognition and is connected with emotions moved by beauty and connection with the artistic energy of the world that created this music and literature. If your person does open up, consider making notes after the session or tape recording sessions (if your person does not mind—don't let the taping intrude).

Notes:

• Consider ending each session with more music for calming closure. During this time you can rest yourself and let the relaxation of the session flow through you and restore you.

• Keep a journal of your activities and what was discussed. This could be a wonderful record for you and other generations after your person has passed on.

• Create a *Special Spiritual Signature* for your person by making a list over time of his/her favorite music, pictures, colors, and readings that may be used to sooth your person at the end of his/her life or used at his/her funeral. This may sound decidedly spiritual, but even if you don't believe in things spiritual you may want to use your person's favorite things to bring them pleasure in his/her last moments in time.

• Read the section on death. It is an important subject that should be included in your discussions at some point.

Follows a certain religion

Your person's religion provides a context for you to participate in ritual and to establish spiritual sessions. The religion/spiritual beliefs a person has followed in his her lifetime contain stepping stones on his/her spiritual path.

• Continue to participate in religious ritual if it brings your person spiritual connection and comfort.

• Collect favorite prayers and music from your religion for your spiritual library.

• Adapt the ideas proposed in the section: **Believes in God, Supreme Beings, or Universal Divine State** (page 165) to reflect yours and your person's religious heritage.

Has a good relationship and would like to talk to a clergyman/woman/spiritual teacher

If there is such a relationship available to your person, ask this person if s/he would meet with your person from time to time for a spiritual conversation or a time of prayer.

• Prep the clergyman/woman/spiritual teacher as to your person's level of disease and ability to discuss religious/spiritual issues.

• Encourage them to have a free and open discussion to the extent your person can participate.

• Have him/her read prayers, bible passages, and offer blessings. This may be the whole session if your person cannot participate too well.

• Attend these sessions with your person unless s/he insists on a private audience. This way you can guide and support your person in expressing his/her desires and avoid confused or inappropriate behavior.

• Use these sessions to identify favorite material for your own sessions.

Does not remember how to behave during religious services

You can act as a gentle guide for your person while attending religious services. This is especially important if your person still enjoys attending.

• Sit next to your person and have other family members sit on the other side. This way you have extra arms if your person tries to get up and exit at the wrong moment.

• Sit in the rear of the church so that you can take your person to the bathroom or outside if s/he makes inappropriate sounds or yells.

Notes:

• Consider the children's soundproof room if making noise is a problem. Tell your person that you both are watching over the children. Sit your person close to the viewing window so that s/he can still see the service. Hopefully there will be a speaker to hear the service.

• Accompany your person to communion if that ritual is desired. Prep the priest/minister to place the host or bread into the your person's hands or yours and then you place it in his/her mouth. If your person has a tendency to spit or dribble, skip the wine if it is a common cup. A small individual serving will be fine.

• Don't worry if your person can't stand, sit, or kneel at the appropriate moments. No one should care. Assist him/her if s/he would like to try; otherwise, encourage calm sitting.

Has no religious or spiritual interest

If your person has entered late stage AD, s/he may no longer be able to verbally or physically respond to any of the material in your spiritual sessions. This does not mean you have to stop. You can continue favorite readings and music. Somewhere these things may be seeping into his/her internal world, which may be consumed with the resolution of life's events, dream landscapes, or making divine connections. We do not truly know. If there is any chance that the sessions in one form or another could continue to aide your person in uniting his/her mortal soul with the immortal one or the Godhead, then it may be worth it. You will have to decide and figure this concept into your belief/value system. You will also have to measure the time you have for this activity in the busy schedule of care.

• Consider music and taped readings. This could lessen your stress and still provide comfort for your person.

• Make these recordings yourself. Your person may enjoy hearing your familiar voice and find it comforting especially if s/he has periods of random irritation.

SPIRITUAL COMMUNICATION
General

Beyond the steps and ideas mentioned in previous sections there could be special issues that arise when you and your person do not share the same religious backgrounds, beliefs, and values. Hopefully it is not a huge or fracturing difference but just a matter of finding common ground. The compassionate compromise may have to come from you the caregiver, as your person with Alzheimer's disease/dementia (AD) may not be able to make the cognitive adjustments.

Person with AD and caregiver share the same religious denomination

This is the easiest place to come from, but there still may be differences over issues of orthodoxy that could have developed over your adult lifetime. Now is not the time to play them up. If you have been the liberal one and your person the conservative or vice versa, time may be on your side. Your person may slowly forget your previous differences.

• Try to minimize differences and work with commonalties. You may have the same appreciation for certain prayers and teachings.

Notes:

• Identify your common beliefs and likes. Spiritual sessions can be formed from these.

• Set aside old hurts and arguments. They will never be settled by confrontation or debate.

• Remember that the idea is to work to the common good.

• Use the suggestions for spiritual sessions outlined in the section: **Believes in God, Supreme Beings, or Universal Divine State** (page 165)

Both are religious but do not share the same denomination

If you never had an argument over your religious differences, then you will simply work toward commonalties. If there were bitter difference in the past then the advice in the above section will be helpful.

• Teach yourself something about the religious denomination of your person. This could be very important if your person is a devout follower.

• Have a conversation with your person as to what s/he truly enjoyed about his/her faith.

• Identify material from your conversation as well as classical prayers and material that your person might like. Include these in your spiritual library.

• Introduce some materials from your religious denomination. Your person may grow to enjoy these items and end up including them in his/her favorites. This way you both can experience enjoyment.

• Remember this is much about what your person might want and need. Be compassionate and set aside your differences. Both of you will benefit spiritually.

• Use the suggestions for spiritual sessions outlined in the section: **Believes in God, Supreme Beings, or Universal Divine State** (page 165)

Person with AD is religious and caregiver is spiritual/ nondenominational

You as the caregiver will be coming from a slightly more liberal background and that should give you a bit more philosophical flexibility. If you are upset about the dogma of a particular religion try to set that aside; you do not have to work with those aspects. Seek to concentrate on more universal beliefs. Hopefully at some point your person will not care so much about strict ideas and enjoy your sharing time together.

• Teach yourself something about the religious denomination of your person. This could be very important if your person is a devout follower.

• Have a conversation with your person as to what s/he truly enjoyed about his/her faith.

• Identify material from your conversation as well as classical prayers and material that your person might like. Include these in your spiritual library.

• Introduce some materials from your spiritual beliefs. Your person may grow to enjoy these items and end up including them in his/her favorites. This way you both can experience enjoyment.

• Remember this is much about what your person might want and need. Be compassionate and set aside your differences. Both of you will benefit spiritually.

Notes:

• Use the suggestions for spiritual sessions outlined in the section: **Believes in God, Supreme Beings, or Universal Divine State** (page 165)

Person with AD is spiritual/nondenominational and caregiver is religious

This may be a more difficult place to come from if your person is uncomfortable with your religious rituals and beliefs. You will still have to be the paragon of compromise and emphasize the universal.

• Have a conversation with your person as to what s/he truly enjoyed about the spiritual.

• Identify material from your conversation as well as classical prayers and material that your person might like. Include these in your spiritual library.

• Look for spiritual materials in bookstores that will satisfy the middle ground for both of you.

• Introduce some materials from your religious denomination. Your person may grow to enjoy these items and end up including them in his/her favorites. This way you both can experience enjoyment.

• Remember this is much about what your person might want and need. Be compassionate and set aside your differences. Both of you will benefit spiritually.

• Use the suggestions for spiritual sessions outlined in the section: **Believes in God, Supreme Beings, or Universal Divine State** (page 165)

DEATH

General

We must all come to terms with our feelings about death or face them at the very last moment of our lives. A number of people cling painfully to life because they have no peace around these issues. Encourage discussions on this issue not just with your person but with the whole family. Much of the first steps are concerned with clarifying values and beliefs. What do you and your person believe in? If you have been working with the spiritual sessions idea, you should have accomplished or be working on this step. The rest is determined in frank discussions and quiet spiritual contemplation. The ultimate peace comes through an individual path.

Expresses feelings about death

If your person will talk freely about death, incorporate it into your spiritual sessions.

• Identify material that will stimulate good conversation and expression of feelings and fears about death.

• Listen openly to what your person has to say.

• Do not try to convince him/her of a particular viewpoint.

• Guide and comfort where it seems appropriate.

• Elicit help from a clergyman/woman/spiritual teacher if either you or your person has one in your life.

• Have these discussions on an ongoing basis even if you do not decide to use the spiritual sessions. This is a big issue for your person and eventually for you.

• Work with the Five Wishes concept. This is a form for recording your person's answers to five major end-of-life questions.

1. The person I want to make care decisions for me when I can't

2. The kind of medical treatment I want or don't want

3. How comfortable I want to be

4. How I want people to treat me

5. What I want my loved ones to know

These forms are often available through your local hospice organizations or can be obtained through (www.agingwithdignity.org). This form is easy to use and is a great record to have to remind you of your person's desires.

Very worried or scared about death

It would be very helpful to encourage frank discussions to assist in getting, if possible, to peace on the issue of death. If your person is reluctant to talk, approach the issue slowly and gently. Follow the above suggestions for talking about death.

Upset over deaths of friends or relatives

This creates an excellent natural time to discuss the issues of death. Use all aspects of your person's feelings for frank discussions. Follow the above suggestions for talking about death.

Will not or cannot talk about death

There may be little you can do, but do not give up hope the subject may open up suddenly.

• Approach the subject in a round-about fashion.

• Pick some material for your sessions see section: **Believes in God, Supreme Beings, or Universal Divine State** (page 166) that softly approaches the subject.

• Take advantage of any natural opening such as those produced by the death of a friend, relative, or even a pet.

• If your person is in too late a stage of AD for spiritual discussions, pray routinely for him/her (if it is in your belief system). This loving energy and intention can help to bring peace to your person.

You have the opportunity to be present at the death of your person.

Death is one of the very real spiritual and physical aspects of living a human life. Modern life has in many ways discouraged people from participating in the sacredness of birth and death. Being present at the death of another human being can be one of the most profoundly affecting moments in your life. It is nothing to fear and can help you bring some peace into your own life.

• Make sure you have the support of loving family and the help of professionals (either in the hospital or hospice).

• Ask for chaplain support if you are in a hospital settings.

• Ask for chaplain or social work support if you are in a nursing home setting. You may be able to get hospice support in some nursing homes.

• Maintain a loving approach and state of feeling.

• Open your heart and experience the love you have for your person. It

Notes:

may feel painful and too emotional at first but you are connecting with a universal divine state of love and the interconnectedness of all living things.

• Go with the flow of your feelings. Stay centered in your own personal spiritual connections. In this place you may feel the equanimity of love and can balance your feelings of loss.

• Concentrate on freedom for the soul of your person. In your silent prayers, give your person your loving permission to pass on. Sometimes the dying cannot leave because the energy of the living is both loving and possessive.

• Carry out any rituals that your person requested.

• Surround your person with elements of his/her spiritual signature (see explanation in section on spiritual sessions) as appropriate.

• Take guidance and support from Hospice nurses or pastoral counselors if they are available.

• Hold your person's hand or sit by the bed and stay in loving contact and connectedness until your person has died. Take breaks as needed.

• Allow yourself the room to accept death in any way that it happens. If you person dies while you step out for a break, just accept that it needed to happen this way.

• Take time to heal. Expect to feel an open state of sadness that can send you to tears at a moment's notice. Live in this sadness don't drive it underground. Slowly it will pass and you will be left with a feeling of deep loving connectedness with your person.

• You will have the satisfaction of knowing that you gave one of the greatest spiritual gifts of compassion-being a caregiver of someone who could not care for him/herself.

• Many blessings will fall upon your soul for your selfless act of love.

CHAPTER 4
ALZHEIMER'S DISEASE

A. BASIC DEFINITION

It is the intention of this book to provide you with basic information to get you started on the road to problem solving and dealing with the job ahead. Everyone who deals with this disease needs to understand that Alzheimer's disease is:

- A chronic brain impairment

- Global: eventually impacting the whole brain

- A disease with no definite starting point

- Slowly progressive and (at this time) irreversible

- Usually active in the brain long before symptoms appear

- A form of Dementia (a category of disease)

- Characterized by deterioration of cognitive (thinking brain) functions: speech, abstract thought, memory, social abilities, emotion, and finally the ability to care for oneself

- Seen in the brain as plaques crowding around the nerve cells and tangles inside, which eventually cause the destruction of nerves and causes the brain to shrink or atrophy

- A disease with no currently known cause

- A disease that slowly removes adult abilities moving your person backward to childlike skills (**retrogenesis** – reverse development).

BRAIN CHANGES: The brain is an amazing and complex organ. Its deterioration is traumatic and dramatic. The changes that occur help to explain why a person can no longer function in an adult way. The brain is made up of millions of nerve cells. As we grow from infancy these nerve cells become covered in a protective sheath (myelin). This sheath serves to turn them on and strengthens their signals. The first area to become sheathed is the motor area that allows us to move our arms and legs. Next the areas to be developed allow us to feel touch sensations (parietal region), to begin to see (occipital region), and to hear (temporal region). The brain slowly builds from these early skills to a greater and greater understanding of the world around us. We begin to develop language and the ability to concentrate, learn, think, and plan. Through a lifetime of learning our frontal lobes become the powerful seat of our thinking and intelligence. The last area of the brain to be sheathed in myelin is the hippocampus (assists in generating memory) and the amygdala (regulates emotions). This brings us to the beginning point of Alzheimer's disease. The last to be protected is the first to be lost. Scientists have learned from examining the brains of persons with Alzheimer's disease that the plaques and tangles start in the area of the hippocampus and the amygdala. This is why the first symptoms are often memory loss and emotional mood swings. Slowly these plaques and tangles spread to connecting areas mangling and destroying nerve cells as they go. The whole building process is destroyed in reverse. The temporal lobes deteriorate bringing difficulty with language and possibly hallucinations. The destruction of the parietal lobes brings disruption to the control of limbs and sensory input through touch. The person with Alzheimer's disease veiled by confusion, gradually separates from his or her world. As the frontal lobes are affected, the ability to synthesize information and make sense of life unravels. Life becomes a string of isolated and long-held remembrances. No new memories can form, and life begins anew everyday. As each specific skill is lost your person with

Alzheimer's disease becomes more childlike. Planning your care strategies around this fact can reduce many difficulties.

B. DIAGNOSIS

The diagnosis of Alzheimer's disease/dementia involves deciding if the symptoms presented could be caused by any other factors, many of, which are treatable. There are currently no truly accurate medical tests to diagnose Alzheimer's disease/dementia (dementia is a more global term used when your MD is not sure if it is the Alzheimer's type). General dementia and Alzheimer's disease behave in essentially the same ways – the losses are eventually the same and the timing is always individual. A thorough medical exam and mental testing are needed to be safe in arriving at a diagnosis of probable Alzheimer's disease or dementia. Brain scans can be helpful in showing areas of decreased tissue volume as well as blood and glucose flow.

C. SYMPTOMS

Enough has been printed on Alzheimer's disease/dementia to make most adults aware of and nervous about their memory. As people normally age, they often experience some minor difficulty in remembering the names of others as well as a few of life's details. Our modern lives are certainly filled with complexities, making an organized life a challenge for everyone. This is not the memory loss seen in people with Alzheimer's disease/dementia. In this disease the memory loss is progressive and clearly begins to impact daily living, often creating some safety crises. Social skills begin to diminish and there can be sudden outbursts of emotion. The appearance, number, and progression of these symptoms are individual. You have to compare your person's past and present skills to look for major differences. If your person has had problems in a certain area all of his or her life, then it may be not be a valid symptom. People with Alzheimer's disease usually show a group of problems. The following is a list of some of the signs of Alzheimer's disease:

- Forgetfulness or memory loss
- Failure to recognize familiar people
- Inability to account for the day's activities
- Problems with finding appropriate words
- Difficulty with simple math, often causing crises with checkbooks and money
- Difficulty reading and writing
- Displays of poor judgment
- Deterioration in appearance and grooming
- Changes in personality
- Major swings in emotions and mood
- Becoming fearful or anxious for no apparent reason
- Uninhibited behavior
- Disorientation
- Problems with motor coordination
- Getting lost
- Safety problems such as leaving appliance on or pots burning on the stove

• Difficulty driving or getting lost driving in the car

• Telling strange stories that may be hallucinations

It is a natural instinct for people to compensate for their losses and cover them up for as long as possible. Most of us do not want to see ourselves failing at the basic activities of daily living. This sometimes makes it difficult for family members to become aware of a person's problems until a crisis appears. Do not feel guilty if your loved one gets into trouble and then suddenly there is a scramble to get a diagnosis. This happens to many families.

D. DISEASE STAGES

Theorists have struggled to define levels or stages in Alzheimer's disease. The length of the disease varies and medical practitioners like to be able to determine where a person is in the progression of the disease. This can be helpful in planning care. Dr. Barry Reisburg has set up the most accepted staging model with seven major levels and a number of sub-levels. When you are a caregiver of a person with Alzheimer's disease all these stages naturally meld into three: *early, middle, and late.* In the *early* stage, your person still has many self-care skills. He or she may just need supervision of difficult skills. People sometimes still live alone. This is the best time to plan for the future. Where will your person live? Should he or she stop driving? What activities need supervising? What are the safety issues?

Things will worsen and in the *middle* stage and most people need constant support and supervision. This is the longest stage and the most challenging (it feels like *middle madness* to many caregivers). As your person's abilities deteriorate, more and more needs to be done for him/her. During this time your person becomes more childlike. This is the long mile for caregivers. This is also when you need to develop creative problem-solving skills.

In the *late* stage, persons with Alzheimer's disease/dementia will have lost so many skills that s/he will take on an infant-like appearance. The person can no longer talk or walk and minimally responds to the outside world. Care provi sion at this stage is intense (24 hours per day, 7 days per week). Most home caregivers turn over care to a nursing home at this stage if they have not done so toward the end of the *middle* stage.

Eventually, your person will not be able to swallow and the decision whether or not to begin artificial nutrition must be made. Some people die of other health problems before reaching this state. If kept in this state, a person may assume fetal postures, truly reversing his/her development to the infancy. Eventually a person will weaken and die of opportunistic disease or body systems failure. The end is both a sad and thankful event in the progression of Alzheimer's disease. Both the caregiver and the sufferer are released.

CHAPTER 5
THE WHOLE PERSON CONCEPT AND
LIVING WELL WITH CHRONIC ILLNESS

A. WHOLE PERSON CONCEPT

It sounds like a simple idea " the whole person concept " and it is. But it's one that medical disciplines are slow to adopt. Most of the time you hear the concept used as an assist in getting to a high level of wellness. How does this youthful lingo apply to a chronic illness like Alzheimer's disease/dementia?

If you examine all the areas of the whole person (mental, physical, emotional, social, and spiritual) and try to preserve the strengths in them, then your person with Alzheimer's disease/dementia can often be "well" even with a chronic disease. "Wellness" is more than the absence of illness. It is a general sense of well-being and can be felt by a chronically ill person, most especially in the absence of acute symptoms and pain. A "well" person with Alzheimer's disease/dementia is easier to care for than one with complications. It is true that the person with Alzheimer's disease/dementia will gradually deteriorate no matter what you do; but the process can be slowed in some cases, allowing you more quality time with your person.

Generally, all skills are lost in the order in which they were gained. It is like reverse growth and development. An adult with Alzheimer's disease gradually becomes a teenager and then regresses through childhood into infancy. The skills one has had the longest are the last to go. Most caregivers sense the childlike dilemmas when their person with Alzheimer's disease/dementia can no longer dress him/herself (independent dressing is usually learned at 2-5 yrs. of age). This concept helps to explain the gradual decline and childlike behavior. If given time and no other diseases interfere, a person will progress to infantile behavior and even assume a fetal posture. At this juncture your person would have to be fed artificially through a tube in his or her nose or stomach. This theory of **retro-genesis** – reverse development can help caregivers understand the problems they face and some of the appropriate strategies to use to manage behavior. One of the largest dilemmas is how to use strategies that one might use on a child but at the same time preserve his/her dignity. You can overcome this problem by using basic encouragement and support that helps to work with the strengths and skills that each person has until their strengths and skills are lost.

A person in the *middle* period of Alzheimer's will often have strengths in each of whole person areas—strengths that you can work with. This does not have to become a total devotion, but some thought and planning can make your life easier. The following paragraphs will talk about strengths and weaknesses in each area and hopefully give you something to think about.

B. MENTAL

We all know that a person with Alzheimer's disease/dementia has many losses in this area. The person with Alzheimer's disease/dementia has a lot of short-term memory loss, but often can remember the distant past. During the *middle madness* period your person can often obey certain one-step requests but cannot usually "think" him/herself out of problems. The person with Alzheimer's disease needs to be watched carefully, but it is good to let him/her be busy for periods during the day. Give your person one-step commands to do small chores/activities that make him/her feel useful (there are a lot of suggestions in the main problem solving section of this book). As the disease progresses your person will be able to do fewer and fewer chores/activities and for shorter periods of time. It is worth continuing to encourage him/her in the highest level of function until your frustrations draw some natural lines.

C. PHYSICAL

In this area, losses often appear in the later stages of the disease as your person with Alzheimer's disease/dementia begins to loose the ability to walk, talk, and eat. Until that time you often have a fairly spry person on your hands who has little or no common sense. There is no harm in encouraging walking. It can be a soothing activity for persons with Alzheimer's disease/dementia (possibly stimulating endorphins or natural painkillers). Escorted walks are best, but keep an eye on your person even in a locked yard. Adult daycare centers can be a good source of supervised walks and mild exercise.

A word should be said here for encouraging good nutrition. You don't have to go on hyper-nutritional vitamin diet to get health gains. Since people with Alzheimer's disease usually cannot cook for themselves, you have a big say over what they are offered to eat. Getting your person with Alzheimer's disease/ dementia to eat at all can be a battle royal, but a simple moderate-to-low fat diet with fiber can slow weight loss and help alleviate constipation and chronic bowel problems (sweetened oatmeal can be a miracle food). Pay attention to your person's natural hunger cycles. Some like eating breakfast and have dwindling appetites as night falls. Take advantage of these patterns to encourage consumption of more wholesome calories. Remember this is not a cure so don't create manias for yourself. Check with your doctor or nutritionist for diet regimes for specific illnesses. The main text of this workbook has problem-solving ideas for dwindling appetites and eating skills.

D. EMOTIONAL

Alzheimer's disease/dementia can create mood changes and emotions that don't match the circumstances. The person with Alzheimer's disease can laugh when he/she should be sad and vice versa. One can cannot predict when this will happen. The person with Alzheimer's disease/dementia can get into some intense short-term anxieties and agitation, but memory loss can sometimes prevent a person with Alzheimer's disease/dementia from consistently remembering chronic worries. This in turn can sometimes alleviate flare-ups of emotionally related diseases such as ulcers. Remember that you are not always the cause of his/her emotional upsets. S/he can make up the death of a friend or relative and cry over it without any thought or understanding of the truth. S/he will fill in the blanks of their lives with new stories often believable to outsiders. People in general find it very hard to live with blanks in their lives. Filling in the blanks is a natural process that is greatly exaggerated in Alzheimer's disease/dementia. This process is called confabulation (making up stories to explain lost memories) and it can be quite entertaining and/or enraging depending on your mental health and attitude at the time.

E. SOCIAL

Depending on how social a person with Alzheimer's disease/dementia is, his/her basic social skills can stay intact for quite a long time. The key here is stimulation. If s/he has no one to talk or listen to s/he can become disinterested and uncommunicative. In truth it is not always easy or enjoyable to converse with a person with Alzheimer's disease/dementia, but s/he will often settle for light chitchat. Here again, daycare can be a valuable resource. Daycare assistants often enjoy the challenge of talking to persons with dementia (these assistants can escape at the end of the day). A day at the center can satisfy a person with Alzheimer's disease/dementia's need for socialization with peers and take the pressure off you as a caregiver.

F. SPIRITUAL

This is indeed a challenging and often overlooked area that does not have to be limited to organized religion. A person with Alzheimer's disease/dementia has spiritual depths as all people do and may grow closer to the divine as s/he becomes less attached to the reality defined in this world. If your person's communication skills are still intact, s/he can discuss thoughts and feelings about his/her spiritual beliefs. Discussion can help a demented person through the natural stages of mourning over chronic illness and death to create a bit of peace in his/her soul before all communication is lost and the intense march toward death begins. Conflicted feelings and lack of peace can prolong the death process.

Encourage your person with Alzheimer's disease/dementia to engage his/her spiritual belief as a resource. You may need to find a cooperative minister, priest, or spiritual director who would like the challenge of communicating with a person with dementia. Do not be put off by the loss of memory; effects can be created in the moment that may add to your person's inner peace in ways no outsider will ever know. Spiritual feeling can continue even after communication is marred. Church, spiritual rituals, and prayers can sometimes be helpful in sustaining these feelings. For others, ritual is meaningless and a simple expression of feelings by someone who cares can be enough. Verbal and non-verbal communication of love is a powerful tool. Refer to the section on setting up **spiritual sessions** (page 166) for more guidance.

CHAPTER 6
BEHAVIORAL PROBLEM SOLVING vs.
THE MEDICAL APPROACH

A. CONCENTRATING ON BEHAVIORS AND SKILLS vs. THE DISEASE

Alzheimer's disease/dementia at this time is an incurable disease. Everyone hopes that the future has a cure but for now reality is the adventure. Certainly if the person you care for is in the early stages of the disease, exploring treatments to slow down the progress can be important. There are a number of drug treatment regimes going on each year. If the disease is approaching middle stages, it can be a difficult decision. You can treat a person at this stage and sort of freeze him/her in a time of difficult behavior. With a number of drugs, after the treatment is stopped your person returns to the state s/he would have been had s/he not been on the drug in the first place. After the choice for drug treatment and symptom management is made, you are still left with the question of how to live your life.

This is a major shifting moment for caregivers and the persons being cared for. At this specific moment the behaviors and symptoms of the disease are about as good as they are going to get. The next step is to begin structuring and planning your life. The behaviors and skills of the person you are caring for are the place to start. In working with skills and behaviors, you do not just look at deficits or losses but at what is still functional and work to preserve those skills. This will be what much of caregiving will be all about for as long as you are physically involved in the caregiving process. A person with Alzheimer's disease/dementia who is working to his/her strengths is an easier person to care for. Just as a mother celebrates, as a child is able to do more and more things for him or herself, a caregiver needs to allow and promote as much independent behavior as possible for as long as this behavior lasts. The secret of much of this is to keep things SIMPLE. Remove as much clutter as possible and approach tasks in simple steps. The main body of this workbook is filled with ideas to make the SIMPLE possible. This does necessitate slowing down a bit to think things through and to plan. Many things will just take more time. It is almost like living in a Zen monastery where all steps are contemplated. A slower world is a more peaceful world.

B. BEHAVIORAL/SKILL PROBLEM SOLVING

Keeping a problem solving approach is one of your best protections against caregiver burnout. It doesn't matter how long you plan to be the physical caregiver. In the best of circumstances, you want to work in harmony with the inevitable changes in behavior as much of the time as possible. This is not to say that this or any other approach will be a miracle. You will experience some frustration and anger. When you live with a person with Alzheimer's disease/dementia, much of caregiving will be trial and error. You are both individuals with personalities, preferences, and histories. No one activity or idea works for everyone all the time. You need to accept what does not work, and move on to a new idea. Even ideas that work well for a while will lose their use as the behavior changes or the skill is lost.

Problem solving is a systematic way to approach the development of ideas for trial and error. By using the system presented in this workbook you can use the ideas to trial a possible solution and record whether it worked or not. The book can be your memory when things get tough or you are tired and just can't remember what you have tried. It also helps you anticipate what the next skill loss may be and to plan ways to adapt what is currently working to meet the next change. A behavior and skill focused problem solving approach can help keep you out of chaos. There are also benefits to your self-esteem. Everyone wants to feel capable and on top of their problems. You cannot always exert control on the world, but you do have control over your attitudes and choices. You can be in the driver's seat even if the road is rough, as long as you have a good car. The problem-solving approach is that car.

DISTANCE CAREGIVERS NOTE: This approach can also be helpful if you are a distance caregiver, but in that circumstance you do not have a great deal of control. The best idea is for the caregiver that is physically present in the home to have a copy of the book as well, and then the two of you can brainstorm together. You then can be a better support and can even purchase special items, which might be needed in a specific, circumstance and send them as a gift/contribution. This idea does not have to be limited to two family members or friends. A whole network of people can work together with everyone keeping track of the changes and problems with their workbooks. It would also be key if a family were trading the person's care from home to home. One central workbook could be passed with the person with Alzheimer's disease and each care group could have their own as a reference to use for brainstorming and planning.

C. PRIORITIZING WHAT IS IMPORTANT

The next step after accepting the problem-solving approach is to decide what you need to focus on first. This can be a challenge depending on how long you have been caregiving and at what stage of Alzheimer's disease/dementia the person you care for is in. Despite any pressures that you may feel you are under, it is a good idea to have read the "how to" section of the book and perform an assessment of your person with Alzheimer's disease/dementia using the **Behaviors List**. As you go down the list and mark off each level of skill, you will develop the full picture of what stage your person is in and how much loss he or she has suffered. After doing this overview, you can then focus on the problem areas that are bothering you the most and disrupting your routine. Read the corresponding problem solving sections and pick some ideas to try. It is best to work on only one or two areas at a time. When you get the most disruptive problems under control then move on to smaller things or work on preserving the good skills that your person currently has. The following are a few things to take into consideration in each stage of Alzheimer's disease/dementia:

EARLY STAGE: More people are getting diagnosed while many of their skills are still intact. These people need to participate in their own planning and stay independent in their skills for as long as they safely can. If you are a caregiver of a person at this stage, the problem solving system can work for you as well. At this stage you work first with any major losses and work to support existing functions. This is when reminder-notes are helpful. Break actions down into simple steps. Make things easy to find or see, especially tools needed for hygiene and appearance. Dignity is very important. Work with the person you are caring for to plan care strategies. Often s/he will know what will work best. This approach preserves dignity and self-esteem for all involved.

PLAN AHEAD. Take care of financial matters. Consult a lawyer and set up the kinds of "powers of attorney" you will need to handle financial and health decisions. Many of these documents need to be done while your person with Alzheimer's disease/dementia still has "decisional capacity" or his/her wits about them. Work through any trust issues about who should handle things, especially before your person develops any tendency toward paranoia (this does not always happen but can be very troublesome when it does).

If your person has been living alone, plan the next living arrangement before it becomes a safety crisis. If you both are already in your primary residence begin to make a plan to safety-proof (dementia proof) the environment (just like you would for toddlers keeping poisons and dangerous tools locked up). You don't want to do this so soon that you frustrate your person with many skills intact. A plan can be sufficient. You can safety-proof things from a physical function standpoint. Remove scatter rugs if your person has trouble walking and could slip. Equipment can be installed such as shower and tub bars and chairs, as well as mobile shower sprays for easier bathing.

MIDDLE STAGE: As more major skills are lost, you have to switch gears. Personal safety and protection of assets (if not already taken care of) need to be done as a first priority. Implement your safety plan as soon as you see your person with Alzheimer's disease starting to misunderstand the use of soaps, caustic agents, and medicines that could be harmful if swallowed or taken inappropriately. This will be a challenging time, with new problems appearing fairly regularly. The **Behaviors List** section of the workbook can be invaluable to you during this time.

LATE/END STAGE: Care at this stage is moving toward total support of every function a person has. Persons at this stage will need to be fully bathed, dressed, and fed. Most caregivers tire before this stage and place their loved one in an assisted living, shelter care, or nursing home. Complete 24-hour care is overwhelming for most people and you will need hired help or committed family help to take a shift. Home health and other community care programs can be helpful. You may also need the help of a social worker to sort out what you qualify for (see the section on long term care for more information).

D. FOCUS AND CREATIVITY

Once you have decided what you would like to focus on, mark those pages with post-it flags or bookmarks for quick access. Circle ideas you would like to try. Make notes in the margins or in the note section. This is truly a "workbook": make it work for you. Keep in mind one special rule: *Your creativity is a vital resource in all of this trial and error/problem solving.* You are the only person who knows what is happening in your situation. Your friends may come up with good ideas, as well as people in your support group, but you will be the one who has to make the final decisions. This book was designed to compile as many suggestions as are out there in the public domain. Ideas are plentiful but need to be adapted to every new situation. No book is the letter of the law. Adaptation and creativity are the name of the game.

Another key aspect is that of enjoyment. If a problem-solving idea or approach seems like fun to you, try it. All this analysis does not have to be dry and "medical." Making your ideas enjoyable can save you from much wear and tear. As a person with Alzheimer's disease/dementia becomes more childlike, so can your approaches. If you like to walk at the zoo, get a yearly membership for the both of you, and go for weekly outings. Work with your strengths, pleasures, and hobbies in addition to those of the person you are caring for. Your life enjoyment will greatly improve. **Use Your Creativity!**

CHAPTER 7
CARE MANAGEMENT AND WORKING WITH THE MEDICAL PROFESSION

A. WHAT IS CARE MANAGEMENT?

The term derives from CASE MANAGEMENT, which describes what nurses and social workers do when they assist a person in managing his/her health care. These professionals provide oversight and assistance in selecting, accessing and evaluating services, and sometimes they provide direct services. You may have had some more negative contact with case management when dealing with insurance companies who sometimes use this term to define their process of "gate keeping" or rationing out health care services. True case management is a process of working with a person to help them get the services s/he needs. This all sounds very medical and it is. It is important for any caregiver to know what is happening in the broader medical world. CARE MANAGEMENT is simply the process of knowing how to represent yourself and the person you are caring for as you maneuver in the world of medical and social services.

B. CARE MANAGEMENT AND CAREGIVING

If you have been caregiving for any length of time, you have had some experience in dealing with health care professionals, obtaining services, and tracking medical information. If you have had these experiences, you have also been angry, frustrated, amazed and overwhelmed. People will offer advice in every sector, but there is little help for keeping all of it straight. CARE MANAGEMENT is the art of keeping it all straight and more easily dealing with the medical and social systems.

C. THE STEPS OF CARE MANAGEMENT

1. GET TO KNOW THE BASICS OF ALZHEIMER'S DISEASE/DEMENTIA. Learn the symptoms, progress, and problems you will face, treatments offered, and about the difficult decisions you may have to make. Knowing these things will give you a common understanding with medical and social professionals. All this basic information is covered in this book.

2. KEEP A RECORD OF MEDICAL INFORMATION. In a notebook or special logbook keep a written copy of the medical facts about your person with Alzheimer's disease/dementia including the following:

- Name, address and other descriptive information
- Other medical facts: height, weight, usual blood pressure, usual pulse, and blood type.
- Medical conditions
- Past illnesses and emergency treatments received
- Current medications and drug allergies
- Hospitalizations and surgeries
- Note any abnormal lab or test results that you know about

3. KEEP A LOG OF HEALTH CARE PROVIDERS AND CONTACTS. In the same notebook or special logbook keep names, date of contact, addresses, phone numbers and specialty information on the following:

GENERAL PROVIDERS: List all the general providers you are using: physicians, case managers, consultants, pharmacies, medical equipment companies, home care organizations, day cares, homemaker and private duty nursing services. Be sure to keep the names of any nurses, home health aides, therapists, social workers, homemakers and day care workers that regularly work with you as well as their supervisors—this will help you if you have to call an organization about information or problems.

EMERGENCY PROVIDERS: Keep a list of emergency providers and phone numbers that you may need to use: emergency rooms, ambulance-fire department-police services, poison control and mental health crisis intervention services.

COMMUNITY SERVICES: List community services that you are using or may need in the future: The local chapter of the Alzheimer's and Related Diseases Association, support group leaders, supportive members, Meals on Wheels, and respite care groups. In this section leave some space for listing the names of persons you have contacted in these organizations, when they were contacted, what was said or promised, what prices were quoted, and deadlines that were agreed upon.

This probably seems like a lot of information to get down on paper, but every new medical person will want to know about your person with Alzheimer's disease/dementia's medical history. You will find it impossible to keep it all in your memory. Keeping all the suggested phone numbers and contact persons in a notebook will save you the stress of having to endlessly pore through the pages of phone books and scraps of paper. You will also be spared the pain of having forgotten key people you need to contact and the frustration of trying to make a case (with no information or facts to back you up) for services promised to you but not delivered.

D. TIPS FOR CARE MANAGEMENT

The steps for care management are basically twofold: become knowledgeable and keep a logbook. These actions will help you keep your work straight but there are some further suggestions to help you become more skilled:

1. PICK YOUR PROVIDERS CAREFULLY. Your dealings with the health care system will be smoother if you pick knowledgeable physicians and service providers with compatible views. Consider the following:

• Do they have expertise with or have they treated a number of persons with Alzheimer's disease/dementia?

• Do they have a kind, caring manner and are they sensitive to your concerns?

• Do they have compatible views on any treatment issues that you may have?
Will they extend you the time to talk about issues?

In an ideal world all health care professionals should be operating from an area of expertise, with a kind, caring manner and give you the time to have discussions on issues of importance and concern. Health care professionals are trained to this ideal, but some fail to put it into practice. Avoid these providers, life is too short and your life's task too difficult for you to put up with too many people who are not meeting your needs. This does not necessarily give you license to be rude and overbearing with your providers. Operate from an area of strength:

• Let your providers clearly know what you need.

• Listen to what they offer.

• Observe what is provided.

• Give accurate feedback on the good as well as bad service.

• Keep the dialogue up.

• If they are rude or uncaring and/or fail to correct the problems after a couple of attempts, ask to speak to a supervisor,

• If you have appealed to several administrative levels and the problems are not solved and you are not hopeful that they will be, remove yourself from their service. Also lodge a written complaint to the service. If they hope to stay in business, someone in that company will listen and make it better for the next person. Service organizations are in general very aware of the problems that can be caused by one person telling twenty of their friends how bad the service was.

2. KEEP AN OPEN DIALOGUE WITH YOUR PROVIDERS.

The health care professionals that you have chosen can be of great help to you. Keep the dialogue open, bring up issues, and seek their advice and instruction. This does not mean that you have to accept every suggestion. You are the captain of your own ship. A professional may not remind you of this fact, but they are trained to recognize and honor your position as caregiver. In an ideal world health care professionals should assist you to care for yourself and support your caring for your person with Alzheimer's disease/dementia. Unfortunately, many well-meaning but over-zealous professionals will just provide the needed care and may not provide any instruction. If you do not seek or accept teaching you will find that you are expected to provide care with little or no knowledge on how to do it properly. Assert yourself; let them help you to help yourself.

3. ESTABLISH POSITIVE COMMUNICATION

• Expect a caring and helpful attitude as well as quality care, and let the provider know if you get anything less.

• Be friendly and cooperative. This creates a positive working relationship.

• Be knowledgeable but open to new ideas. This is the dynamic of a positive working relationship.

• Recognize disagreement as a constructive process. Make sure your expectations and point of view are known.

Listen to the other's information and point of view. This keeps the discussion open and clear. The rest is a learning experience for both parties.

• Establish your boundaries. This is especially important if you are accepting help into your home (i.e. private duty nurses, aides and homemakers). Privacy is important, but well-meaning care providers cannot always gauge correctly if they have invaded some of your private space. Let them know the boundaries and in a firm but friendly way let them know promptly when the have crossed them.

4. THINK A BIT AHEAD AND ANTICIPATE SOME OF YOUR NEXT PROBLEMS.

Thinking ahead is simply a good planning idea. You cannot anticipate everything nor do you want to indulge in catastrophic thinking. You should not always look for the worst scenario but simply the next one. By using your behavior list and marking current abilities, you can see which ability may be lost or which behavioral problem may surface next. Read ahead to problem solving strategies for these anticipated situations. This activity will 1) give you some time to think about the suggested coping strategies, 2) tentatively decide what may work best for you and 3) think of creative adaptations to the listed strategies.

5. LINE UP YOUR RESOURCES BEFORE BIG CHANGES.

This can be difficult because some big changes are sudden (i.e. heart attacks or emergency surgeries). Develop some emergency plans. This will make the burden of decision making in emergency situations much less stressful. If the occasion is a planned surgery or changes in the environment, ask your health care providers what to anticipate, and line up the help you may need before you make the change.

6. NEVER UNDERESTIMATE THE IMPORTANCE OF YOUR SUPPORT NETWORK.

You cannot provide effective care for a person with Alzheimer's disease/dementia for very long if you are socially isolated. This isolation can increase your sense of burden, which may lead to neglect of yourself and/or the person you are caring for. There is unfortunately an identified trend toward poor health in caregivers. This may be due to either the age of the caregiver and/or the level of burden and stress s/he is experiencing when providing the care. The job of caregiving is never easy but trying to do it with little or no help or support is foolish. You have a better chance of maintaining your own health if you marshal all the resources available to help you with the challenge of caregiving. **CHAPTER 12** contains more information on building a support network.

CHAPTER 8
COPING PERSPECTIVES FOR CAREGIVING

A. RETROGENESIS:

Retrogenesis is a relatively new term coined by Dr. Barry Reisburg after many years of observing and investigating Alzheimer's disease/dementia. The technical definition is: "the process by which degenerative mechanisms reverse in the order of acquisition in normal development."[1] With retrogenesis the last skills that one learned, generally will be the first to go. The person with Alzheimer's disease/dementia is going backward in time in a pattern of **retrogenesis** – reverse development. Those who have dealt with Alzheimer's disease/dementia for any length of time will intuitively recognize this to be the truth, and it sets the stage for improved problem-solving strategies. Most of us are ingrained to accept adult behavior from adult-looking persons, but persons with Alzheimer's disease/dementia present a special dilemma. A person is prone to react in anger when an adult suddenly does something childlike. Reisberg has set the stage for a better understanding of Alzheimer's disease/dementia. As a caregiver, you need to examine the difficult behaviors and problems you are trying to handle in this new light. What is the mental age (retro-age) of your person? How old is s/he acting? Is this a behavior you might see in a two-year-old? Does this emotional reaction seem like a temper tantrum? The answers to all these questions and others like them should be considered before developing an attitude toward your person and his/her disease. Without this knowledge, many caregivers have reacted as if all the erratic behaviors of your person with Alzheimer's disease/dementia were directed at them personally (getting back, revenge or torture). It is best if you try to seek to understand what is motivating your person's behavior (i.e. a temper tantrum for a snack or confusion over what to do next—with faint remembrances floating around in his/her head). Caregivers have a large task. They have to keep the person they are caring for safe, fed, and clothed in a shifting world that they have to make some sense of or play along with. As a caregiver, I often felt that I was Alice in Wonderland trying to catch up with the red queen.

B. LIVING IN THE SHIFTING WORLD OF ALZHEIMER'S DISEASE

As you can see, having the world shift backward in time is a strange slice of reality for you and the person you are caring for. Viewing the changes as a somewhat orderly pattern of decline can be very helpful in managing this changing world. It is true that each person's changes will be determined by his/her own biology and disease; but you can begin to understand what will change next. Flexibility and creativity are your main mental resources. Try not to cling stiffly to your own view of reality. If possible try to step into the world of the person you are caring for. It does not have to be 5 PM if it is 8 AM in his/her world. In past times, medical personnel would obsess over "re-orienting" your person with Alzheimer's disease/ dementia making sure s/he was informed of the real date and time. No visions or alternative realities were acceptable. This "re-orienting" process caused a great deal of anguish and created behavior problems. Over time this idea was abandoned for persons with Alzheimer's disease/dementia. Today most professionals understand that "going with the flow" is the answer. Some people are afraid that to deny the true date and time or to pretend you are living in the past is lying. They valiantly keep trying to insist on reality, but this strategy will inevitably fail. If you are one of these people, you will need to become more flexible or you will not be able to survive as a caregiver. When you get into the flow with the shifting worlds, it can at times be lighthearted and even fun. You do not have to hunt bears with the person you are caring for, but you can allow them to. If you are trying to get something accomplished, let them go on with their adventure, especially if they seem to be enjoying themselves and are not in any danger. If the person you are caring for has hit a sad memory and is caught up in it, then distraction is a great idea. Depending on where s/he is in his/her regression, food or a special activity can keep him/her occupied until the old memory begins to fade. The object is peaceful co-existence and some degree of happiness for all. Be creative and work with the flow.

C. PERSONALITY AND ATTITUDE:

In these times, more and more people are coming forward with troubling symptoms and being given the diagnosis of Alzheimer's disease/dementia. These people can be physically healthy and have time for introspection. The generation/ group preceding them was captured in shock and shame in their decline. It was a traumatic event to come forward, and many still react in protective denial. Out of our knowledge from these individuals with early diagnosis have come ideas about trying to live well with the illness and the hope that doing some attitudinal work early on may help mitigate later behavior and problems and emotional outbursts. No one is sure if this will work but it does seem that your person with Alzheimer's disease/dementia's previous personality will naturally play into the grace or turmoil of his/her decline. If you think about returning to childhood, you can imagine the pleasant aspects of seeing the world through eyes of wonder just as a child does. If your person's childhood was a dark place filled with abuse, neglect or emotional starvation, you can imagine that returning might be emotionally difficult. It would be lovely to think that a person like this can have a pleasant regression, but there is not enough information to be firmly predictive. As caregivers, you may be the only people who understand your person well enough to know when your person with Alzheimer's disease/dementia is entering some difficult childhood memory and have the beneficence and human connection to lead him/her out of it. Hope, love, and trial and error are the best resources in dealing with past and present personality issues. If you think there are serious problems and you are not able to deal with the emotional and behavioral fallout, then consulting your physician or a psychiatrist may be necessary. In other cases, losing memory can be a good thing. Stress illnesses like ulcers may clear up because your person can no longer hold onto long-term stress issues. All these issues are individual and you will have to take into account your person's adult personality as well as his/her skills and talents in trying to understand his/her decline in Alzheimer's disease/dementia and how to manage it.

Attitude can be everything in the shifting world of Alzheimer's disease/dementia. Anger is often an immediate response to behavior that you consider difficult or even outrageous. Knowledge that your person is in regression and cannot help some of the problems s/he gets into can go a long way to mitigating your anger. Understanding and problem solving are your best antidotes (a sense of humor is a great help as well). For example: you come into the room to find the person you are caring for has just dumped most of a bottle of furniture polish on a table. Your flash reaction might be to be angry and lash out at your person, verbally chastising him/her severely for making such a mess. Your anger can trail on as you snatch up the bottle and run madly to find rags or towels to mop up the mess. The person you are caring for will most likely stand there staring at you, unable to comprehend the fuss. S/he may feel that it was just a helpful job. If you stop for a second and think about the situation, it is not much different than raising children. Four-to-five-year-olds love to help their parents and imitate their behaviors. A young child could easily get into a scrape like this. As a parent you may be a bit angry, but you know s/he doesn't know any better. It is not truly different for a person with Alzheimer's disease/dementia. S/he looks like an adult, but his/her mental and often physical skills are that of a child. There is also the question of dignity. No one enjoys or thrives on humiliation—

neither adults nor children. Mistakes are made. At this juncture you can change your attitude and make the best of a difficult situation. First you can take a moment to be thankful that s/he did not drink the furniture polish necessitating a trip to the ER. At that point you would know your person's retro-age was closer to two. Being thankful for small blessings, you move on to gently letting your person know a mistake has been made and a clean up is needed. You remove what's left of the polish (just in case your person wants to finish off the job with a drink) and proceed to quickly get some towels or rags for the polish that is headed for your carpet. You return with the rags and enlist the aid of your person in the clean up, which most likely will consist of slopping up the excess and rubbing the rest into the table. As you both rub away, the crisis slowly ebbs away and no one has lost life or limb. If your person can't get into table rubbing, don't fret. Let him/her play around or get him/her to fold the rags (which can mean moving them in and out of the basket or some variation thereof). *Note: a basket of rags, light broom, or small push sweeper are good, handy tools to keep around for clean up and to use as a distraction or for redirection when you don't want your person in your way. Even in a demented state, a person wants to feel useful and included. If you have not raised children you will have to use your imagination but for those who have, this is a very recognizable problem. Your reaction to problems needs to be gentle. You can be firm with the behavioral limits, but you need to redirect and distract.

D. MULTIPLE INTELLIGENCE

Many of us were raised in the time when IQ was the great predictor of our future. Parents lamented lower scores and accepted lack on this single measure of intelligence as a sign of low abilities. Times have changed. Howard Gardner, an educational theorist, brought us out of this darkness and broadened our definition of intelligence from one single measure to a number of distinct types.[2] When one examines Gardner's list of types of intelligence, one realizes how logical and intuitive it is. Every parent has seen their child bloom in some special way and recognized that it would never be accounted for, measured, or valued in a single system of testing. How might this pertain to elderly persons with Alzheimer's disease/ dementia?

As stated previously, a person's talents and skills need to be taken into account when deciding how to manage his/her behaviors. Encouraging positive behaviors and happy experiences can make caregiving a much easier task. Even as a person declines, his/her natural talents and proclivities can be sustained. Some will love to work with art materials and others with math or puzzles. Picking activities for which your person has no talent or desire can provoke negative behavioral reactions. Examining the person you are caring for in this light can bring greater understanding to your caregiving.

The following are the areas of "multiple intelligence" as defined by Howard Gardner:[3]

MUSICAL INTELLIGENCE: The center for making music is located in the right side of the brain. How much this talent will be affected will depend on when it manifested and how much training your person had. If music was a love in the life of the person you are caring for, work to preserve it. Delicate musical instruments may have to be removed because they are no longer safe, but a fake or toy may bring comfort. Just watching or listening to others play may be important to his/her happiness. This is where caregiver knowledge and perception is important. You will know when an activity is soothing or upsetting. Having a special talent that you can no longer practice can bring on grief. Use your creativity to find those things that bring joy verses sorrow.

BODILY-KINESTHETIC INTELLIGENCE: Exceptional use of the body can be seen in athletes, dancers and even inventors. In the brain these talents are governed in the motor cortex and often driven by your "handedness"- left or right (right handedness is controlled by the left side of the brain). If your person has been gifted with these talents, work with what is remaining. Tossing balls can be engaging or watching sports on TV. If you do not care much for sports; but the person you are caring for does, use your network to get someone to volunteer to take your person to watch a game or to toss ball in the park. If your person was a dancer or a body artist, his/her memorabilia may be important (such as costumes, pictures, or music). Watching dances could be enjoyable. When it comes to mementos be careful if the items are valuable; your person can go into a period of forgetting and destroy a precious item. It is better to remove the originals and give them replicas or copies that won't matter if they are destroyed. For inventors this can be very important, since the only tangible results of their talents may be objects.

LOGICAL-MATHEMATICAL INTELLIGENCE: Along with language skills, this form of reasoning construes what we accept as IQ. It is also seen as scientific thinking. Persons with this type of intelligence are rapid problem-solvers; ideas can just pop into their heads. If the person you are caring for has been a scientist, a mathematician, or a teacher with excellent logic skills, you will want to find ways to stimulate this talent. It can be simple, perhaps a notebook of figures and calculations. Nothing has to go into it, but they can imitate their old skills and find it involving and comforting. Some people love jigsaw puzzles. You can make these puzzles ever simpler by buying them in the toy store and matching them to your person with Alzheimer's disease/dementia's retro-age. In all these circumstances you will have to use your own creativity, taking cues from your person.

LINGUISTIC INTELLIGENCE: This kind of intelligence can be severely attacked by the brain damage from Alzheimer's disease/dementia. Most persons as they regress will have difficulty retrieving words and forming correct sentences. Poets and writers have linguistic talents. Again mimicking a beloved activity can bring comfort. Emerson used to review his journals and remark on their cleverness, even when he did not know who wrote them.

SPATIAL INTELLIGENCE: Artists of all varieties work with spatial problem solving. The right side of the brain governs spatial intelligence. In recent years a number of books have been written about stimulating your artistic skills by exercising the right side of your brain. Navigators, chess players, architects, and people who work with their hands creating furniture or fixing machines would also have spatial talents. Take your cues from the person you are caring for. Objects from the past, three-dimensional puzzles or a small set of blocks (i.e. Frank Lloyd Wright's freuble blocks might be an idea, though they are expensive) could be useful to occupy the spatial mind. It is fairly easy to create activities for people with artistic talents. There are numerous kinds of art notebooks, paints, and pencils that are safe to use. Remember to keep it simple so that you aren't constantly running to find lost items. You can train your person to return the art objects to you when the activity is over, by giving them gentle repeated reminders and cues. This could save you trouble from the "hiding" problem. You could also have several sets and when one disappears you have another to use until the first one turns back up again.

INTERPERSONAL INTELLIGENCE: People with this kind of intelligence have the ability to "read" another person's intentions and desires. This skill is very intuitive and is often seen in political and religious leaders, psychotherapists, parents, teachers, and nurses—people who have a talent for working closely with others. This talent is governed by the frontal lobes of the brain and is not as quickly damaged by Alzheimer's disease/dementia, as are other areas. This is why many people with Alzheimer's disease/dementia maintain their outward appearance and cursory social skills long after their ability to reason and actually hold a coherent conversation has gone. People with interpersonal skills are attracted to being around other people. They may enjoy day care where there can be lots of opportunity for socializing, dances, and group activities. Just as pre-school is valuable for children, day care can be a wonderful asset for most persons with Alzheimer's disease/dementia and give you a needed break.

INTRAPERSONAL INTELLIGENCE: This form of intelligence is less tangible than the other forms. It is the talent and ability to access one's own internal life at a deep level. Understanding one's own emotions and how they relate to behavior helps to build a vital, creative being. This type of intelligence is essentially private and needs to be expressed to be detected (via artistic expression—language, writing, or music). If you are caring for a person who has this kind of intelligence, perhaps s/he kept journals or writings; these can be mimicked for a long time. Stimulating writing and reading (real or imagined) may be the best way to work with this talent.

E. COPING WITH GRIEF

When you first read about grief, it seems like a fairly contained process. A person in your family dies, you experience deep sadness, and then you feel better. Actually there are stages of feeling, and people experience them differently depending on their culture and life circumstances. Some theories say these stages come in a certain order, but practically speaking they can happen in any order. The most important fact is that you can experience grief even if the person you care for is not dead but has a chronic disease or disability. Parents of children with disabilities experience grief and it is continuous. Now this may sound excessively gloomy, but caregivers learn to cope. Grief tends to twinge us depending on the current losses of your person. This is why this long process is called "living with grief." If you learn to recognize and accept grief, it can be a healing process.

A number of authors have developed feeling states of grieving. The following is a brief description of these states as they might apply to caregivers of persons with Alzheimer's disease/dementia.[4] Remember grief states may happen in a different order, some may not be felt, and they can come and go.

NUMBNESS AND SHOCK: This stage often comes with diagnosis. Most families suspect something is wrong but the label "Alzheimer's" or "Dementia" holds a special sadness and relief. You finally know what is wrong. The search is over, but the future is uncertain. You are not sure what you feel or what to do about it. This is a difficult stage for decision-making. You need some recovery time.

DENIAL: It is hard to accept a chronic and terminal diagnosis. Even when the medical evidence seems overwhelming, you grapple for hope. You need time to adjust. Sometimes you react to the label "Alzheimer's" or "Dementia," worrying that your person with the disease will be looked down upon by other people and that it will reflect on you. People can often indulge in blaming the victim: "She caused this to happen to herself." It's easy to see why you might not want to jump right into accepting all these problems. Take time to sort it out.

ANXIETY AND FEAR: The future with Alzheimer's disease/dementia is very uncertain. As a caregiver, many burdens seem to loom on the horizon. Anxiety and fear are natural reactions to this powerful change in your life. There is nothing wrong with experiencing the feelings and then letting them go as you sort out ways to handle the future. Excessive anxiety, which interferes with your ability to deal with daily life, can be a problem. There is help from support groups to counseling. Discuss the problem with your physician. This anxiety and fear is probably temporary and does not indicate "mental illness." Experiencing emotions and examining them is internal "work" and is a necessary part of life and mental health. If suppressed, these feelings can cause problems like chronic anxiety and illness. Support from friends, family, and professionals can help over time to put life in perspective again.

GUILT: Talking a person out of their guilt is an impossible task, but guilt is a fairly useless emotion and very destructive if you slop around in it. Sure a twinge of guilt can support a healthy conscience, but only a twinge! No matter who you are, husband, wife, son, daughter, etc.; you did not cause this disease. Perhaps you have not treated your person with Alzheimer's disease/dementia in the best way all your life together; but this disease is not retribution. Try to leave the guilt behind! You do need to make peace with yourself, sort out your feelings about your loved one, and decide how you will approach your future relationship.

DEPRESSION: In many ways your loved one has been lost to Alzheimer's disease/dementia. Sadness over loss is natural. Depression comes in various doses and levels. We all experience moments of it in our lives. You still have time to be with your person with Alzheimer's disease/dementia and in fact many years may lie ahead of you. This can make you thankful, especially if you were very close, or it can seem scary and promote depression. Either way, if you feel yourself sinking into depression, try to be nice to yourself. It's a good time to get some respite care, even for a few hours. See a funny movie; laughter is good medicine. Caring for a person with Alzheimer's disease is a long committed task and for many of us there is little escape. The best idea is to build on your strengths and capitalize on any happiness, large or small. Try not to isolate yourself from the world; it just accentuates depression. If you feel that your depressive feelings are not lifting or are becoming chronic, seek medical advice. Medication and counseling can be effective.

ANGER: In my opinion, it would be a miracle if you escape this emotion. Few people accept life's adversities without resentment and anger. You can still love your person, while resenting the burden s/he may represent in your life. Anger can often be the sum and expression of depression and sadness. You can watch your person stumble through the house on some futile quest, be overwhelmed with loss of your person they once were, and then allow yourself to get caught up in some angry argument (that you might otherwise have avoided). Emotions can run in unruly bunches not in the neat little ways that are described in books. Anger can appear when you least expect it or want it. Give yourself time out if you can. If you are worried about excessive anger and/or fear that you might harm your person, seek help. Remember that a demented person can easily provoke you, especially if he or she may be in a mean, snooty disposition. Unfortunately persons with Alzheimer's disease/dementia cannot be truly held accountable for their actions like other adults. Think about their "retro-age", what childlike level is being displayed? You do have the resource of setting behavioral limits and letting them know when you have to step away for a bit to cool off. Persons with Alzheimer's disease/dementia can sense emotions even when words are not understood.

EMPTINESS, LONELINESS, AND ISOLATION: These feelings can be experienced when you are alone as the caregiver. The person with Alzheimer's disease/dementia has been diagnosed and some of the fuss has died down. It may then hit you that there is a long road ahead. Some of your friends may drift away because you are no longer as available for social activities. You may suddenly feel isolated and alone. This is not unexpected. So much loss has been experienced. You need to allow yourself to feel the grief. Isolation and withdrawal from social activities can create a safe space to regroup your thoughts. After a time, then make plans to get out when you can. You can meet new friends through support groups or day care. Reach out to others when you can. If people offer to help, don't turn them down. Invite them into your support network, even if it is only a small offer of help. All the help builds up.

BODY REACTIONS: Your whole body grieves a loss. Finding out that someone you love has such a difficult debilitating disease can create an energetic shock to your body. It could manifest in increased headaches, appetite loss, or increase in emotional eating patterns, fatigue, or indigestion. If you have such a somatic reaction to your grief, accept it and treat your symptoms as needed. This is a reactive state and it will likely pass. If symptoms persist, see your physician. You may need some counseling to get you through this time. Do not be ashamed of needing some extra support to work out your feelings. Some of the most successful clinics and physician practices have counselors on staff to give support after diagnosis. All of us struggle through these states, but everyone can benefit from the support. Talking to close friends about your feelings can be very helpful to ease the distress. A good friend should be included in your support network (see **CHAPTER 12**).

Adjusting to all this loss is very difficult and it is ongoing. In fairness, there is truly no gain in painting a rosy picture of caregiving for a person with Alzheimer's disease/dementia. Professionals do steer away from the negative, wanting always to create pretty pictures of love, kindness, and hugs. I have been personally frustrated by this rose-colored view of caregiving. Love and hugs are wonderful, but sometimes that will not be how you feel. Don't let other people's vision of what is NORMAL and COPING depress and anger you. Acknowledge your feelings. I recommend a good hard cry now and then. Play music that makes you let it all out. Have hope. You will learn to cope and survive with creativity and problem-solving techniques. Find what works for you—your own path.

NOTE FOR "SANDWICHED" FAMILIES: If as a caregiver, you have children and a person with Alzheimer's disease/dementia all under the same roof, be alerted to the fact that all of you are shared caregivers and will share in the continuous grief. Individually you may each be in a different feeling state, which makes for interesting if not stressed interactions at times. It can also give you something to talk about. Take time to let family members talk about their feelings, especially their angers. Bottled anger is your worst enemy, particularly with teenagers. As a group you can let it all hang out in these conversations, laugh a lot, and share all the crazy stories. One of our family favorites is when my teenage son found his grandmother, who had Alzheimer's disease, in the middle of the road. Worried that a car might hit her, he quickly brought her home. When he got over his shock, he asked her why she had been standing in the road! She briefly explained that she was trying to stop a car to take her to church. My son responded " Grandma you shouldn't be hitchhiking. It's not safe!" She quickly retorted, "It's not hitchhiking if it's in the name of the Lord!"

Professional counseling can be helpful in the beginning, to help the children and adults learn to express their grief and anger constructively. You don't have to use counseling forever, but it can build good communication skills and social support for your whole family. Another resource is adult day care. Our kids worked there as volunteers and paid summer staff. They found acceptance for their feelings and pride in their skills working with persons with Alzheimer's disease/ dementia. You can search your community for these services, as well as support groups. Just do not let yourself or your family become socially isolated. It will turn caregiving into a lonely and tortured experience.

In the everyday swing of things, rotating chores within the family can help to relieve the family member who is in the most emotional distress. For instance, I took my mother-in-law to the beauty parlor for a perm. We picked up groceries on the way home. She decided she was hungry in the store and began a pestering campaign to get food. This was supposed to be a short stop and we were to have lunch at home. I offered her some meat and cheese samples that she rudely turned down. I told her she would get a peanut butter sandwich when we got home. She loudly proclaimed at the checkout that she hated peanut butter and why did we always give it to her. The looks from the crowd were mixed. In my worst moment I imagined that they all envisioned me locking her in the closet feeding her endless PBJ. On the other hand a large number of people noticed her wandering in circles and figured something was wrong. I finally offered her a banana. She ate it, after loudly proclaiming it wasn't ripe enough. By the time I got home, I was worn out from her rude and demanding behavior. Her retro-age was 5-6 years old. At home I had my oldest daughter feed her and I took a rest. Interestingly enough in the end, she happily ate a peanut butter and jelly sandwich (She was offered ham but turned it down). Middle Madness at its best.

Taking time out can be the best medicine. Don't feel guilty if you have had enough. If you are a lone caregiver make sure your person with Alzheimer's disease is safe, and you go to another room for a breather.

F. STRESS RELIEF

YOU ARE A VERY IMPORTANT PERSON- TAKE CARE OF YOURSELF.

Your own health is a vital key to your success as a caregiver. As you become more knowledgeable and skillful in your role, some stressors fall away; but there will always be stress. You need to take stock of your own resources:

- What is your current level of health?

- Do you have special health problems?

- Are you performing the activities recommended by your healthcare providers to keep yourself healthy?

- Are you getting enough rest and relaxation?

- Are you still keeping up with activities/hobbies that provide you with fun and enjoyment?

Hopefully these questions will make you think about how you are taking care of yourself. If you have concerns about how you are tolerating the stress of caregiving, talk to your physician and go in for an exam and consultation. There are also many books on stress reduction and relaxation at most local bookstores. In your caregiver's notebook section make some notations and plans for maintaining and improving your health.

Sometimes providers are quick to recommend getting away from it all on a frequent basis. Trips away can be difficult to negotiate and stressful in their own right. If a big break works for you, take advantage of it. But little breaks are even more important, grabbing quiet moments every day can let negative attitudes and issues roll away. What is better for you? Getting all the clothes folded while the person you are caring for naps, or taking a few quiet moments to relax or meditate with your eyes closed? You will have to be the judge. But remember, as humans we are often driven to destruction by our own expectations and our own imagining about the importance of what others think about what we are doing.

Another important factor is to keep your sense of humor; it is a vital way to keep a healthy attitude toward what you are doing. Even in the worst of situations there can be a glimmer of humor. Laughter stimulates endorphins, natural painkillers and mood elevators. A chuckle or a guffaw can be one of Mother Nature's best gifts.

CHAPTER 9
RELATIONSHIP DILEMMAS IN CAREGIVING

Relationships are always a challenge and life's vital connection. Alzheimer's disease/dementia changes the timber of all the relationships that come in contact with it. Everyone has to adjust. This adjustment does not come without work. If diagnosed in the early stage, persons with Alzheimer's disease/dementia have a small window of time to actively participate in adjusting relationships with their loved ones. After a time, the changes in their mental abilities and emotional responses generally prevent them from being very concerned about the emotional lives of others. They begin to regress and lose the complex adult understanding of relationships. Retrogenesis gradually brings them to a more childlike view of the people in their lives. In middle stages, persons with Alzheimer's disease/dementia often have intermittent recognition of who people are. If they are not sure, they will fill in the blank and assume or fit that person into a familiar slot. After my mother- in-law was living in a shelter care, she no longer recognized me as her caregiver and daughter-in-law. I became a friend from work, and the owner of the shelter care took over my identity. This can be a very disconcerting and a hurtful aspect of the shifting world of a person with Alzheimer's disease/dementia. You have to emotionally and mentally adjust to this "reality" and stay flexible. Once things start changing and your person "confabulates" or makes up a new reality to fill in the memory gaps. The confabulation may not stop. You may be someone new each time your person sees you. If you go along, it can be a social exchange. If you try to correct him/her-to make yourself feel more comfortable, it may provoke a negative and emotionally out-of-control response. It is not worth it! You will spend much time repairing the damage, calming your person down, and feeling very stressed about the whole encounter. Following are discussions of some special relationship dilemmas.

A. SPOUSES AND SIGNIFICANT OTHERS

There is no question that this is one of life's great tragedies. No matter how many years you have spent living with someone you love, discovering s/he has Alzheimer's disease/dementia is over-whelming. Alzheimer's/dementia is now a well-known disease, and people know that accepting the challenge of caregiving means a commitment measured in years. It is not like cancer where your person may be in severe physical decline and pain, but s/he generally remains her/himself. In Alzheimer's disease/dementia, a caregiver experiences waves of loss as his/her loved one regresses back to childhood. As a spouse or significant other (SO) caregiver, the indignity can overwhelm you. I once heard a spouse say: " If only he could see himself, he would never make these choices." How true, but your person will only see him/herself for fleeting moments and will never again see the big picture or be able to rationally correct behavior. Statements like these reflect the deep crisis of adjusting. You have to let go of the person you knew and form a different relationship with a person in flux. This is not easy.

If you are a spousal or significant other (SO) caregiver, be forgiving of yourself. You need to be vigilantly aware of your reactions to problems and the need for adjustment. If you become angry or depressed and let this flow into your caregiving relationship, it could be a problem. Persons with Alzheimer's disease/dementia are fragile people and can be emotionally abused. If you are worried about your ability to adjust and cope with your feelings ask your MD for a referral to counseling. You will not have to go counseling forever. Many people need short-term counseling to adjust to tragedy and loss in their lives. There is no shame in asking for help. If you commit to home caregiving, you may be involved intimately in this relationship for a number of years. It may be unendurable if you are in deep personal distress. Seek help and support.

You can find companionship and sharing in a support group. Many of these caregivers are experiencing the same relationship challenges and riding the same emotional roller coaster. Use your "special/understanding" friends (not everyone will understand and some will be frightened—leading to rejection) and support network to help unload stressed emotions. Read **CHAPTER 12, on Accepting and Building Support** to create a circle of caring people.

The human spirit is amazingly resilient and people learn to cope with incredible suffering, tragedy, and pain. We pick ourselves up and go on. But if caregiving is not for you, then find other ways to be loving and supportive of your spouse or SO wherever s/he ends up living.

B. ADULT CHILDREN

With Alzheimer's disease/dementia no type of relationship is spared adjustment. Adult children have seen their parents from a certain changing perspective. As little children, you are dependent and trust that they will protect and love you. This reality may not have been fulfilled. In an adult view, the parents may not have lived up to childhood desires and needs. Or perhaps they did and you are deeply attached. We all have parental agendas for love and approval, no matter how we were treated. Even abused children will initially make excuses for their parents out of their overwhelming need for love. Again you have to be aware of your responses and work through an adjustment process. It will never serve you to continue to look for parental love and approval from a person with Alzheimer's disease/dementia. This disease often robs people of their ability to empathize with and to even understand the needs of others. Like small children, persons with Alzheimer's disease/ dementia become concerned with themselves and how their own needs will be met. The bigger picture of relationships remains in his/her story but not in active life. S/he can remember the antics of little Joey, but may have no clue that you are that person. Your parent will eventually forget who you are. At that time s/he will no longer be able to make long-term memory connections, the past becomes everything. In his/her eyes, you will fluctuate between a young person and a child. Your parent will lose the memory and recognition of you as you are now. You will probably always be familiar to your parent even if s/he is not sure who you really are. You have to let your love ride with this shifting "reality" and be whoever s/he thinks you are. It is the same for all Alzheimer's disease/dementia relationships, correcting your person will not be helpful.

If your relationship with your parent was very negative or abusive, you should seek counseling before you decide to be a home caregiver for your parent. You may not be able to overcome the problem and acting out within the relationship could be harmful for the both of you.

C. DAUGHTERS-IN-LAW AND SONS-IN-LAW

This role is becoming more of a reality as families become caregiving units for parents with Alzheimer's disease/dementia. Often a parent is single, divorced or widowed and can no longer live alone. It is a large act of love for a family with children to take in a parent with Alzheimer's disease/ dementia, but it can also be a bonus as there are more caregivers to share the work. The adult child may have strong desires to care for his/her parent. The spouse may go along in an effort to please. It will still be a great challenge for the "in-law." If you are in this situation, you will have to examine your own relationship with the parent with Alzheimer's disease/dementia. Whether you are the daughter-in-law or son-in-law, you are accepting a large burden that will change your life. This can be especially tricky if you have small children. Again we all adapt. Children can gain a great deal working with a grandparent with Alzheimer's disease/dementia. They will be raised understanding empathy and compassion. As the "in-law" you will have your measure to give and you will have to forge a commitment to caregiving. Your role will depend on whether you are the primary breadwinner, share that role, or are the primary caretaker. These roles are somewhat more flexible and less gender based than they used to be. Whoever is the primary caretaker will have the larger burden and may face the biggest adjustment. There are wonderful gains in caring for another person's parent. Your acts of compassionate caregiving can open your heart, as well as build the character and expand the loving nature of your children.

D. ADOLESCENTS

When adult children decide to take a parent with Alzheimer's disease/dementia into their home, there can be adolescent children still living in the home. Adolescents are in a flux period themselves, trying to figure out how they fit into the world. A demented grandparent can be shocking at first. Adolescents are certainly aware that their grandparent is no longer the same, but the regressive behavior of Alzheimer's disease/dementia violates the logical world or how things are supposed to be. If the former relationship with the grandparent was close and loving, an adolescent will feel great loss and grief, just as their parents will. Some adolescents will adjust well and accept the shifting world of Alzheimer's disease/ dementia as if it were science fiction or fantasy. This will depend on how attached they are to the "correct" world. Children still have some flexibility in their thinking and imagination that may be harder for their parents to attain. Other adolescents may act out, fearing a loss of attention, love and concern from their parents. Individual or family counseling may be needed to put relationships into perspective. Families who accept caregiving can benefit from short-term group counseling. This can help the group coalesce into a caregiving unit that supports its members well. Adolescents, like other individuals, can gain holistically from being part of a caregiving unit.

Parents do need to be sensitive about the volume and type of duties given to children and adolescents. Toileting is best supervised by adults until it is clear that older adolescents can cope with this task (This can be a sensitive area for both the grandparent and the child). You will have to use your best judgment keeping in mind the developmental stages of all the members of the family. As your parent becomes more childlike, the company of children may seem quite natural. Children/adolescents can share playful activities with a grandparent. If done with the proper supervision, it can be a great relief to you as a caregiver.

E. FRIENDS

This is an unusual circumstance but possible. If you are a friend and have volunteered to care for a person with Alzheimer's disease/dementia, you have committed to a great act of charity. You are probably in the best position to be somewhat impartial about the disease, personality, and behavioral changes. Virginia Bell and David Troxell in their new book *A Dignified Life* speaks of the best friends' approach to care. This book gives you caregiving ideas from the tolerant perspective of a "best friend." Even relatives can adopt this view, because this is where you find yourself once all your previous agendas are cleared. Being a true friend to a person with AD is perhaps the best perspective.

CHAPTER 10
A SPIRITUAL PERSPECTIVE

A. SPIRITUALITY AND ALZHEIMER'S DISEASE

Alzheimer's disease/dementia is a condition, which stimulates deep fascination and enfolds deep sorrow. No one who has ever had it touch his or her life, doubts this potential. It contains a circle of being, often ending life as it began, in fetal form. The Omega is the Alpha. No other disease embodies this developmental regression of the self. It is a disease that strikes great fear in our overworked, over-achieving culture. No one wants to envision himself/herself without all his/her faculties, and even worse reduced to a babbling child in an adult body. This is not the romantic inner child of the self-help world.

There is one major area of questioning that stands in the shadows and has only recently begun to be explored. What happens to the spiritual consciousness of a person with Alzheimer's disease/dementia? It is possible that consciousness may be subject to the same developmental regression? In Alzheimer's disease/dementia there could be a rich inner life that is available while the exterior life is committed to involution (slowly folding in on itself). These possibilities point to the classic human search for meaning— the antidote to emotional pain and suffering. To capture this meaning is to create hope.

Alzheimer's disease/dementia does produce altered states of consciousness, and all of these states are not likely to be merely passive periods of confused brain function, but possibly periods of some receptivity and perception.[1] Could spiritual activity or growth be taking place? Carolyn Myss has postulated that: "In some of the cases of these types of disorders, the individual withdraws his or her normal consciousness in order to have access to processes of spiritual development that we as yet know nothing about. And thus, we only can assume that the state of consciousness your person is in is dysfunctional and seemingly without 'purpose' as we understand purpose."[2]

Currently, Dr. Elmer Green has done the most groundbreaking exploration of the possibilities of spiritual consciousness in Alzheimer's disease/dementia. In his three-volume work entitled *The Ozawkie Book of the Dead*, he discusses his experiences caring for his wife.[3] He chronicled his observations of her behaviors and her states of mind. His conclusion is that Alzheimer's disease/dementia is a **Bardo** state.

The most familiar state of the **Bardo** is the **after death** realm in Tibetan Buddhism. In Tibetan, **Bardo** simply means transition of gap between one situation and another. "'Bar' means 'in between' and 'do' means 'suspended' or 'thrown.'"[4]

When defined this way you can recognize the "in between" nature of the Alzheimer's disease/dementia state. We recognize that death will come, but a person with AD may spend years "suspended" in a childlike state. This disease has been referred to as a "living death" or an "ongoing funeral."[5]

Dr. Elmer Green describes the Alzheimer's disease/dementia consciousness as wavering in and out between " the physical–astral world of your personality and the more-subtle astral world of the soul."[6] This astral or energy body has telepathic and intuitive senses that the physical body lacks. A person with this disease has some advantage over those who are imminently dying because they have a long slow path, with time to become conscious of the "Light of the Soul" or the "higher self " and possibly merge the "mortal self" with it. Obviously this is not an easy path for a person to take; there must be some help. Dr. Elmer Green worked closely with his wife over the time she was consumed with her disease and acted as her guide and support. He read aloud spiritual books to her during her

living Bardo in much the same way the Tibetans read to the dead to guide them through the **after death Bardo.** What this produced, was in Dr. Green's terms, "psychokinesis" in which his wife was able to come out of an infant-like state and speak perfectly, seemingly without the help of her very diseased brain.[7] Her ability to speak seemed to be beyond her brain capacity. Although this could be considered a "moment of lucidity." This lucid state is very rare in the late stages. It may be that as the brain of a person with Alzheimer's disease/dementia is slowly consumed with plaques and tangles, the brain's hold on normal consciousness becomes progressively weaker. Thus a person with Alzheimer's disease/dementia may be able to work at a more holistic level and make spiritual progress that would have ordinarily been difficult.

This could mean that in regression your person returns again to the simple spiritual openness that s/he possibly had as a child. Childhood is a period of fresh receptivity to the world. Children who are not abused believe in the unseen world and sense the interconnectedness of all living things. As much as we fear being forced to return to childlike behavior caused by brain damage, it could be an opportunity for spiritual connection. In many ways reading prayers, reading spiritual texts, and listening to sacred music does not differ from engaging other books and music. It is in fact, all stimulation. Even if your person cannot respond with pleasure, s/he is taking in the sensory energy. In the world of spiritual healing, it is believed that these sounds carry power to affect the body even if there is no mental under-standing. Thus there is no harm in sharing spiritual practice with your person; and it is more likely that it will be beneficial for the both of you. At this point it is not important which spiritual or religious texts, prayers, or music you use. It is important that as the caregiver and reader you enjoy this practice. If you have to force yourself, or have a negative attitude, the practice will most likely not be helpful, calming or healing. The person with Alzheimer's disease/dementia can pick up on your emotions and react to them. The very essence of spirituality is attitude not rote action. A great Christian mystic, Evelyn Underhill believed that spirituality was not a life spent in a monastery but the attitude you hold when performing simple daily tasks.[8] To act as a spiritual guide and a caregiver over the long years of illness is a difficult and dedicated process. No matter what the outcome of bringing spiritual practice into the life you share with your person, there will be few losses and many gentle blessed gains.

If your sense of spirituality means attending church, then it is a good idea to continue unless your person becomes disruptive. If closely guided, your person can still enjoy church rituals such as communion. Whatever your religious beliefs and practices there is much to be gained from keep-ing this part of your life open and healthy.

In honoring spiritual states of consciousness in Alzheimer's disease/dementia, your person can be seen as a human elder with spiritual needs and gifts. In a number of tribal shamanistic cultures around the world, an elder's visions would be incorporated into the cultural milieu and myth. They would be seen as contributing spiritually to the group psyche of the tribe. Our Alzheimer's disease/dementia elders are often put away out of our sight, and the shadow of shame around their "damage" is sustained. As a society we fight over their human status and what should be the humane nature of their treatment. In limiting our views to these struggles, we overlook the still vibrant qualities of their souls.

B. SPIRITUALITY AND CAREGIVING

In taking a new look at spiritual awareness in Alzheimer's disease/dementia, we cannot overlook the special role of caregivers. The devotion of caring for a person with this disease is a spiritual gift and a sacred role. This certainly can be seen in the actions of Dr. Elmer Green. The factors that may distinguish the drudgery of caregiving from its more spiritual form are **compassion** and **commitment**. Many people who have tried to enter "caregiving" in the context of the health professions have been "put off" by having to provide ordinary service to another person. In these cases, money is not enough of a motivator to create willingness to approach what is perceived as servitude and "drudgery." Some people seem to fear this service for its close proximity to enslavement.

Carolyn Myss describes this dilemma of fear in the description of the servant archetype:

"The servant engages aspects of our psyche that call us to make ourselves available to others for the benefit and enhancement of their lives. This task can be done in a healthy manner only if the Servant is simultaneously able to be of service to the self. Without the strength to maintain your own well being, the Servant becomes consumed by the needs of those around you and loses all focus of the value of your own life."[9]

Caregivers who have successfully embraced this service to self and others have done so by making a commitment to the role and to the person they are caring for as well as exercising compassion. In caregiving for a person with Alzheimer's disease/dementia, there is no end of "thankless" jobs and often the care receiver is unable to appreciate the level of sacrifice. The source of the strength to endure must come from within for there is no sustenance in the outward search for thanks and congratulations. Foster in *Celebration of Discipline* talks of the power in the spiritual discipline of service: "Of all the classical Spiritual Disciples, service is the most conducive to the growth of humility. When we set out on a consciously chosen course of action that accents the good of others and is, for the most part, a hidden work, a deep change occurs in our spirits."[10]

It is in the power of choice that commitment is formed. As Foster says "Voluntary servitude is a great joy."[11] In choosing to care for a person with Alzheimer's disease/dementia, you have made a challenging commitment. There will be times of great sorrow and soul searching. Am I doing the right thing for myself and your person I am caring for? The thankless moments can pile up and challenge even the firmest commitment. This is when the practice of compassion for self and others takes you as the caregiver within, to realms of love and forgiveness—the divinity within. In Buddhist tradition this compassion is the heart of **Bodihcitta**. Lama Surya Das describes the dharma or truth of this intention: "Let precious bodhicitta be your organizing principle. Help, and do not harm others. Cultivate these remembrances in everything you do: be gentle, be kind, be thoughtful, be caring, be compassionate, be loving, be fair, be reasonable, be generous to everyone—including yourself."[12]

In Buddhist terms, Dr. Green's dedicated caregiving and spiritual guidance of his wife during the course of her illness falls under this directed compassion or **Bodhicitta**.[13] Rinpoche poetically calls this state of compassion the "wish fulfilling jewel."[14] and Pema Chodron calls it a "tender place."[15] Caregiving for a person with Alzheimer's disease/dementia is truly about **compassion** blended with **commitment**.

In this light, caregiving can be viewed as a spiritual role of service, enhancing your karma (positive or negative energy built up from the consequences of your actions) and self-esteem. Caregiving is a practice and an opportunity. Dr. Rudolph Ballentine in his book *Radical Healing* describes the challenge of looking at the "obstacles" in our lives: "The secret of fulfillment in your life is dealing with the problems that seem to plague you and prevent you from moving forward. Whatever is 'in your way' is your way."[16] This is a deep essence of becoming a caregiver for a person with Alzheimer's disease/dementia. It is a door and a path to a spiritual journey of sacrifice and service. Though personal, this journey needs to be viewed from a noble vantage point and be given spiritual credibility. This is a journey that can bring one closer to the whole of mind, body, and spirit.

The door has been opened to a new view of Alzheimer's disease/dementia consciousness. In this light, caregiving may also be seen in a new way as a transcendent selfless act, which can bring a special dignity to both your person with the disease and you, the caregiver.

C. THE POWER OF PRAYER

One of the simplest ways to keep spirituality alive in your life is prayer. It does not matter what your religious background is; there are prayers in all languages for all the peoples of the world. Foster says: "it is the discipline of prayer that brings us into the deepest and highest work of the human spirit."[17] **Compassion** for self and others is the heart and motivation of prayer. There are times when we know that divine intercession could be beneficial. For a caregiver of a person with Alzheimer's disease/dementia, these times will abound. Small everyday prayers can become the staff of your life, bringing you soulful peace and well-being. It is far better to pray about your difficulties than it is to hold onto anger and negativity. Prayer can help you hold onto **commitment** and **compassion** during your caregiving journey and can become the foundation of a spiritual practice that can support you for the rest of your life. If you choose prayer as one of your spiritual outlets, you will not be alone; in various surveys nearly 75% of caregivers say they use prayer for coping.[18]

Larry Dossey in his landmark book *Healing Words* defines two different types of prayer originating from the *Spindrift Organization*: "directed" and "non-directed."[19] Simply put a "directed " prayer states a specific outcome that is desired and "non-directed" prayer is open-ended. Both of these methods have been shown to work in various studies.[20] You can choose your own method or follow one dictated by a particular faith/religion. The wording of your prayers can also be self-created or you can recite religious or inspirational prayers. In your personal prayers you can share your hopes and fears as well as your gratitude and wonder. As you finish your prayers, take time to let them go and commune in the mystical silence. As author Timothy Freke says in his book *Spiritual Traditions:* "prayer is about using words to bring our awareness to the quiet stillness of God."[21]

D. INSPIRATION AND CONTEMPLATION

If prayer seems too religious or does not fit your views, reading inspirational books and contemplating on the meanings may be a supportive form of spiritual practice. It can also be used with prayer. Inspirational books range from the Bible and other ancient religious texts to contemporary interpretations and thought. Many new books are published each year. If you have never looked for this kind of reading material, just roam around a bookstore and pick something that appeals to you. You can be guided to what you need by your physical and emotional reactions (Read the next section on intuition.). If you feel a certain sense of excitement about reading a particular book, it will probably be a pleasurable and healing experience. This type of reading can transport you some distance from your problems and worries. It is, in a sense meditation in the form of contemplation. You may have to read late at night when the person you care for is asleep. Or perhaps you can just create an afternoon reading break by giving your person something to read at the same time (It does not matter if s/he can read or not, pretending to read can be a comforting ritual). The time you spend thinking about what you have read, no matter how short can be beneficial to the clarity of your thinking and bring you comfort in an often hectic and shifting world.

E. INTUITION

Intuition is simply the **sense** or **instincts** that each of us brings to our decision-making or learning processes. Through work and practice it can become quite powerful, but at its quietest level it is common to all of us. If you honor your intuition, it can become a helpful force in your life. Caregiving for a person with Alzheimer's disease/dementia, is a natural field for intuition. In combination with an understanding of the disease and the types of effective behavioral options, you can use intuition to pick and choose interventions. You know many things about the person you are caring for—some of this knowing is conscious and some unconscious. Intuition combines both types of **knowing** and helps you work more effectively. You may just **sense** something is a good or bad idea. Parents use intuition when caring for children. You often **know** when something isn't quite right and that a child might be doing something s/he shouldn't. Once you become aware of intuition you will notice its effects in your life. As you work with it, intuition can become more powerful and helpful in enhancing your life and creativity.

CHAPTER 11
WORKING WITH ALTERNATIVE OR
COMPLEMENTARY THERAPIES

A. GENERAL SUGGESTIONS

Alternative/Complementary/Energy therapies are ancient methods that work in harmony with the body's natural force or energy. It has only been in recent times that we have come to believe that all healing must come from outside the body in the form of medicines and surgeries. Conventional medicine certainly has its place. Many diseases and conditions do benefit from drugs and surgery, but these modalities are only a small part of the total picture of medicine. The therapies described in this section can be alternatives to conventional medicine, but very often they are complementary and use energy in the body and in the environment to actualize the body's own healing powers. These natural therapies can be useful in treating many diseases including Alzheimer's disease/dementia. There are some general suggestions one should keep in mind when selecting a complementary therapy for a person with Alzheimer's disease/dementia:

• **Do no harm!** The therapies you select should be gentle. A person with Alzheimer's disease/dementia is already in a state of stress. Their world has been turned up side down. At this point, conventional medicine has no cure—just medicines, which attempt to alter the course of the disease. Complementary medicine may act in a similar fashion, but reactions can be quite individual. What works for one person may not work for another. A therapy should be stopped at the first sign of pain, suffering or discomfort. The old adage of "no pain-no gain" does not apply to persons with Alzheimer's disease/dementia. If approached in a beneficent manner with only your person's comfort in mind, positive outcomes are more likely to occur.

• **Keep the therapy simple.** If too many complex tasks, sights, and sounds are occurring at once, it can create sensory overload and a negative reaction can occur. All aspects of a prospective treatment should be relaxing or engaging on a gentle level.

• **All directions must be simple and require little or no memory.** Persons with Alzheimer's disease/ dementia have profound memory impairment. Practitioners should be aware of this, but a caregiver must look out for your person s/he is caring for. If your person can follow simple exercise directions, then that is a bonus. If not, then s/he may still be able to imitate; but just like a child his/her skills may not allow these imitations to be exact. The therapy must be free enough to allow for partial imitation without disruption of the class or else be provided in a private session. The person with Alzheimer's disease/dementia does not want to be laughed at or humiliated for poor performance. This is important when considering exercise, yoga, or Tai Chi classes.

• **Therapies that are applied "to" your person may be the best choices for Middle stage, especially if your person can no longer follow directions.** This can be the kinder, gentler choice if your person is easily confused. At this stage the therapies must be slow and gentle on the senses or your person may become increasingly confused. A person with Alzheimer's disease/dementia can misinterpret actions and touch. If your person becomes agitated by a massage for instance, then that therapy may not be effective for him/her. Another choice is to limit the therapy to one body area, such as back and arms, or legs and feet.

• **There should be pleasure and relaxation involved in the therapy.** Since results of any modality for a person with Alzheimer's disease/dementia cannot be guaranteed, the time involved in the therapy should be immediately rewarding. A relaxing therapy can produce a positive feedback loop and condition a person's behavioral responses. This can prove useful in decreasing stress and agitation. There are modalities that can be done in the home, such as light, sound, music, and gentle massage that can be used to calm a person who is showing signs of high stress.

• **It may be wise to consult a physician who works with complementary therapies.** Doctors of Naturopathy, or Oriental Medicine can also be used. This may be especially important if you wish to try nutrient, homeopathic, or herbal remedies. One can increase vitamin intake to moderate levels without too much difficulty, but mega doses of vitamins, homeopathic medicines and herbs can be very potent. Anything that your person ingests that makes them gag or vomit should be stopped. Even if one believes that ingesting a particular mix of herbs may be a cure, it should do no immediate harm. Doctors who are knowledgeable in complementary therapies can monitor the outcomes of these therapies, changing them as needed to produce some benefit. This can be very stress relieving for a caregiver.

• **Obtain referrals from your physician, national or local credentialing organizations, or from friends who have had successful experiences.** As with any service, cold calling from names in the phone book can have very mixed results. Healthy, caring people should provide complementary therapies in pleasant, positive surroundings. If you do not feel comfortable with a provider, do not go back. You must trust your own intuition, and the provider.

• **Get estimates of the cost.** Insurance companies generally do not pay for complementary therapies unless you have a definite medical condition like muscular trauma, which can be treated or controlled in the short term (Acupuncture is sometimes covered for pain). So the cost of most complementary services will come out of your pocket. With this in mind, you will want to be careful and find out how many treatments are being recommended and at what cost.

• **Don't forget yourself!** If you are looking for complementary therapies for your person, consider them for yourself as well. If a class in gentle yoga works for your person, then you should participate as well. Even if you are busy guiding your person, you will still benefit from any stretches or poses that you manage to do. Massage is wonderful for reducing stress. You may have to get creative and have both massages done in the same room so that you can keep an eye on things. Again your problem-solving skills can be beneficial in this area as well. It is much about a good quality of life for both of you.

B. BODY WORK

This category of therapies deals with physical manipulation. It can be defined as "any system of treatment in which the practitioner uses his or her hands to bring about beneficial changes either in the client's muscles and skeleton, or, through them, in other parts of the body."[1] There are quite a few therapies in this category, but this section will only cover the most popular and most likely to be tolerated by a person with Alzheimer's disease/dementia. There are many books on the market (in general bookstores) that extensively cover the history, use, and possible therapeutic outcomes of each of these therapies. It is a great idea to educate yourself on your favorites.

C. CHIROPRACTIC

This therapy consists of techniques of spinal manipulation. Most people are familiar with its application in back injury or general body trauma. If this is a therapy that has been effective for one or both of you in the past, then it may certainly continue to be valuable. Much attention is placed on alignment of vertebrae. This therapy can be especially helpful if you or the person you care for has back problems.

D. OSTEOPATHY

This technique is similar to chiropractic, but is practiced by Doctors of Osteopathy. These physicians can also practice general medicine and can be very valuable in providing musculoskeletal manipulation and general health care. Osteopaths believe that pain, discomfort, and disease originate in imbalances/upsets in the structural body.[2] Remember the body manipulation needs to be very gentle for a person with Alzheimer's disease/dementia.

E. CRANIAL OSTEOPATHY

Osteopaths practice this therapy as a series of light manipulation to the bones of the skull to enhance the flow and rhythm of cerebrospinal fluid. Some recent studies have suggested that the build-up of toxins such as aluminum and some of the symptoms of Alzheimer's disease/dementia may be caused by sluggish cerebrospinal fluid.[3] Movement of this fluid may bring some benefit.

F. CRANIOSACRAL THERAPY

Massage therapists with advanced training can also provide similar cranial manipulation. The aim of the therapy is the same, but it is usually practiced in the context of massage.

G. MASSAGE

This is one of the most popular complementary therapies. It is an excellent tool to bring relaxation and energetic harmony to the body. You can get a full body massage or simply a massage of various areas of the body: head and neck, back, arms and hands, as well as, legs and feet. Massage is very flexible and therefore can be very effective for a person with Alzheimer's disease/dementia. You can pick and choose what you think will be the most comfortable. Some persons may not understand the need to remove clothing for a full body massage, even though they will be covered in a sheet. In ,this case opt for a more localized massage—feet and hands can be especially soothing. As with everything this will be trial and error, but may be one of the most accessible and easily rewarding complementary therapies. There is much healing power in the simple quality of touch.

H. AROMATHERAPY

The power and pleasure of essential oils from plants and flowers has been known for centuries. Each scent has its characteristic smell and healing properties. You can obtain the positive effects by using these essential oils in a bath, mixed in massage oil, and in atomizers or scent dispensers. Lavender and sweet orange are thought to be calming scents and therefore valuable for people with Alzheimer's disease/dementia.[4] There are many books on essential oils and all of their properties. If you are interested in aromatherapy, do exercise good safety precautions. Open candle flames in dispensers could pose some danger. Also avoid leaving out bottles of these liquids, in case your person might mistakenly drink them. Persons with Alzheimer's disease/dementia do lose their sense of smell, but some theorists believe that the an essential oil does not have to be **smelled** to have value.[5]

I. REFLEXOLOGY

Essentially reflexology is a specialized form of foot massage. It is believed that there are points or areas on the foot that correspond to various organs and areas of the body. Massage of these areas can bring the body back into harmony or balance.[6] The foot is a fairly easy area to massage on a person with Alzheimer's disease/dementia. At its most simple level, this therapy is comfortable and relaxing, thus this may be a more accessible therapy.

J. MOVEMENT THERAPIES: YOGA, T'AI CHI CHUAN, QIGONG

These movement theories come from ancient forms of medical and healing practice. Yoga involves moving into various positions/postures and holding them to get maximum stretch. Yoga is effective in encouraging relaxation and reducing stress. If your person has learned some level of yoga in the past, then taking a class could prove very helpful. If this is a new idea then you will have to find a very tolerant, relaxed class, so that some confusion can be tolerated. Most yoga teachers do not care how long a posture is held; and this allows people to work at their own pace. A gentle system called Restorative Yoga can be quite comfortable for elderly persons.

Tai Chi Chuan and Qigong are oriental systems of movement. This practice of gentle movement promotes the flow of energy or Chi through the body bringing harmony and health. If either you or your person have followed this practice before, it would be wonderful to continue. If you are just starting out, then seek a class for older persons. The directions will be simpler and the pace slower. Both these forms of Chinese therapy are thought to stimulate the brain and help mental agility.[7]

Taking movement classes together can be an enjoyable experience, especially if you suspend fear of embarrassment. For the most part, people taking these classes are trying to enhance their lives with energy balance. Tolerance and caring should be part of this practice. If people treat you poorly, move to another class; and let them own their own feelings and issues. If your person becomes disruptive, then this therapy is not for you. A more solitary activity may be better. Up to this point, working in groups can be of benefit to you and your person on all wellness levels.

K. PHYSICAL EXERCISE

Believe it or not this is a complementary therapy. The body needs movement and exercise. Your heart needs to occasionally beat at higher levels to promote good function. Your physician has undoubtedly told you at some point to get out and walk. This is sage advice. Walking, although not specifically geared toward spiritual enhancement can involve that as well. Walking in your neighborhood, at a nature center or at the zoo, can be a relaxing and uplifting experience. Beauty is everywhere from clouds in the sky, to flowers by the roadside, or leaves blowing in the wind. Fresh air and fresh images enhance your lungs and your soul. This is a simple practice you can do together. Starting up may seem like a problem but after you conquer the logistics, it can become an easy pleasant habit that enhances both of your moods and the day.

L. HERBAL AND HOMEOPATHIC MEDICINE

Over the counter sales of herbal and homeopathic (a form of medicine that believes tiny doses of what caused the condition will alleviate it) preparations have become commonplace. A number of these products advertise enhancement of brain function. Ginkgo Biloba is one of the single preparations that is believed to improve cerebral circulation and mental capacity.[9] Simple dosing of a single preparation is relatively easy, but if you want to experiment with more complex preparations and combinations, it would be wise to work with a doctor who practices Oriental, naturopathic, osteopathic, homeopathic or ayurvedic (a Hindu form) medicine. As with regular medicinal drugs, herbal and homeopathic preparations can be very potent and dosing is tricky. Doctors who utilize these complementary therapies are schooled in the art of creating healthful combinations that could take untrained person hours of trial and error to discover. These specialty doctors can also take your feedback and make alterations to your program. Elderly people are more sensitive to drugs in general and can more easily develop side effects, if the doses are not carefully monitored.

Homeopathic remedies can also be used safely as home remedies for simple ailments (like gas and digestive upset) and first aid (bruising and sprains). Health food stores carry many of these preparations, and can assist you to select appropriate oral and topical (skin) remedies. It is also good to do some reading to understand homeopathy in general.

M. DIETARY AND NUTRITIONAL THERAPIES

A wholesome diet containing healthy quantities of meat, vegetables, fruit, and whole grains is a health booster for everyone at every age. Decreasing excessive amounts of products made with white flour (i.e. bread and pasta), sugar, and fat is also important. This said, getting a frail person with Alzheimer's disease/dementia to eat much at all is a trick. Compromises will be made. Encouraging your person to eat small amounts of the most nutritious foods is obviously a good idea, but supplements and nutrient shakes can be helpful to boost calories and nutrients if needed. Nutrient shakes are a fairly easy idea. There are a number of these nutrient shakes in Health food and regular grocery stores. The taste can be enhanced with egg and ice cream.

Vitamins are a powerful enhancer of health; and because of depleted farming soil, it is hard to get a full compliment of vitamins from normal foods. If your person can and will swallow pills or liquids, then there are some valuable vitamins and minerals that may enhance health and brain function, which can be added to his/her diet including: Coenzyme Q10, B6, B1, B12, folic acid, C, E, Selenium, Magnesium, and Zinc.[10] DHEA, the mother of youth-giving hormones, in the body is found in higher levels in persons who are aging healthfully and is found in very low levels in those who have chronic illness. Some experts think that supplementation with DHEA may have a beneficial effect on many chronic illnesses. In general, dosing decisions can be tricky, especially if you want to use large doses. Consult a doctor who practices Alternative/Complementary medicine to help you make healthy choices and monitor your program.

N. ACUPUNCTURE AND ACUPRESSURE

Acupuncture is an ancient Chinese method that uses very thin needles to promote the flow of energy or Chi or energy through out the body. In this system, disease is thought to be produced by interruptions or blockages in the flow of energy down specific paths or meridians in the body[11] In the western world, Acupuncture has become most well known for its effective treatment of pain. Some believe that scalp acupuncture can be effective in enhancing mental clarity in Alzheimer's disease/ dementia.[12] If your person is in a fairly confused state this may be a difficult therapy to use. Your person may not understand or tolerate the insertion of the needles. If this is the case, it is possible to use Acupressure—a similar system that uses finger pressure instead of needles. A trained practitioner, most likely a Doctor of Oriental Medicine will be needed for these therapies.

O. HEALING TOUCH AND REIKI

Both these practices fall under the category of energy healing and use the hands to pass energy into the body. Healing touch is the more western version of hands on practice and is often taught to nurses as a way of enhancing their practice. Actually "laying on of hands" has been around a long time as a healing remedy.

Reiki is a Japanese therapy and is taught to lay people. Classes are available in many communities and vary in price by the level being taught. Level one is the most basic, least expensive and was taught by the original teachers as a form of energetic home first aid. It is a gentle, soothing and easily practiced therapy (done with clothing on), and is ideal for persons with Alzheimer's disease/dementia. It can be used to treat a myriad of conditions as well as to bring energetic balance and relaxation to the body.

P. SOUND AND MUSIC THERAPY

This is an excellent across-the-board therapy. People can enjoy it at all ages and levels of health. It is very effective in Alzheimer's disease/dementia because the area of the brain that regulates the ability to appreciate music is one of the last areas to be damaged by the disease.[13] Don Campbell was one of the first to publicly promote music as a healing art. He is most famous for his studies of the special healing powers of Mozart.[14] Music may improve sleep, appetite, language skills and socialization in persons with Alzheimer's disease/dementia.[15] Each person is different in their reaction to music and sound. Choices for soothing music range from popular to classical and Gregorian chant. Start with what you know your person already likes and experiment from there. Play music that you find soothing so that you too can benefit from the positive effects of music in the environment.

Q. ART THERAPY

Art therapy is probably as old as the cave paintings and rock carvings. It is not hard to believe that primitive man was looking for inspiration, stimulation, and release in those early representations. Creativity can survive even in disease. The modern artist Willem De Kooning went on to paint for a number of years in spite of Alzheimer's disease/dementia.[16] He expressed his feelings through art even as his mental abilities were declining.

From doodling, to drawing and painting, many people find art relaxing. If your person has displayed artistic talent in the past, s/he should be encouraged to continue unless it is unsafe to do so. Otherwise experiment with simple art endeavors that might be enjoyable, such as drawing with pencils and crayons, or painting with watercolors. Gauge the activity by the abilities of your person and consider their "retro-age" (age created by the affects of retrogenesis.) Products suitable for children are also fine for adults who are at that same "retro-age."

R. RELAXATION THERAPIES: BIOFEEDBACK AND AUTOGENIC TRAINING

Both these therapies require following specific directions and therefore may be most valuable in the early sages of Alzheimer's disease/dementia to enhance and preserve function. Autogenic training consists of following a set of relaxing exercises that induce a state of passive concentration. While in this state the mind can attend to self-healing processes. Once you have learned these techniques you can produce positive effects by short periods of practice in many settings. Autogenic relaxation tapes can be obtained through Self-Health Systems, (see the resource section).

Biofeedback is a method of learning how to consciously regulate body functions such as breathing, heart rate, and blood pressure. Some practitioners believe that mastering this technique may reduce stress and rebalance brain function in Alzheimer's disease/dementia.[18] Biofeedback does require training and the use of specialized devices to feedback the body's vital signs so that you know if you are having the proper effect. For this technique you will need to consult a practitioner for training.

S. CHAKRA HEALING

In Hindu philosophy, the body has seven energy centers or Chakras. They start in your pelvic area and end at the top of your head. Each has a color and an area of power. It is believed that these centers should be open and radiating energy for the body to be in good balance and health. Certain healing practitioners can read Chakras visually or with a pendulum, and can open them. This treatment can be very powerful for health and well-being. People with Alzheimer's disease/ dementia could easily tolerate this therapy. You need to ask around for referrals because all practitioners are not equal. An excellent resource for learning about the importance of Chakras in health and healing is Carolyn Myss' book: *Anatomy of Spirit* (**see SELECTED READINGS** *SPIRITUALITY AND INSPIRATION*). In this book she discusses each Chakra and relates them to Christian Sacraments and levels on the Jewish *Kabalistic Tree of Life*. Once you learn about balancing this energy, you can use visualization and healing hand movements to open your own Chakras as well as those of your person with.

T. ELECTROTHERAPY

This form of therapy is probably most effective for persons in the early stages of Alzheimer's disease/ dementia. After that time a person would probably not understand the use or steps involved in the treatment. Dr. Norman Shealy has created the most useful product and method for this kind of therapy: the Shealy Tens Unit. When used transcranially, Dr. Shealy found that persons with Alzheimer's disease/ dementia had elevation in their seratonin levels and relief from depression.[19] He went on to develop patterns of electro stimulation to acupuncture points called the *Rings of Fire, Air, Earth, Water, and Crystal*. Each of these has been shown to stimulate certain neurochemicals. For Alzheimer's disease/ dementia, he recommends alternating the Rings of Air, Earth, and Crystal one each day for 20 to 40 minutes. The Shealy Tens Unit and Ring protocols can be obtained from Dr. Shealy at Self-Health Systems, (contact information in resource section). You should discuss the use of any healing tools such as this with your physician.

U. MAGNETIC FIELD THERAPY

In the last quarter century much investigation and research has been done on the use of magnets in healing. Dr. Albert Roy Davis noted that positive and negative magnetic polarities effect the body in distinct ways.[20] Positive fields seem to produce stress effects in the human body and negative fields can encourage healing, calm neurons, and encourage relaxation and sleep. As yet there have been no studies to define the specific benefits for Alzheimer's disease/dementia, but therapeutic negative magnets could possibly be used for calming the body and brain.

V. MEDITATION AND VISUALIZATION

Mediation is an ancient practice from the Far East. Although it can be very mystical; it is actually very simple. You do not have to commit to any religious philosophy to begin the practice of meditation. It is more about paying attention. It could be very beneficial to take classes, but the following simple steps can get you started in using a **sitting meditation**:

• Find a comfortable chair. (You do not have to sit in the lotus position to meditate).

• Place your feet flat on the floor, your hands resting in your lap, your back straight and head bent slightly forward.

• Concentrate on your breath moving in and out of your lungs.

• Thoughts will intrude. This is not a problem. Do not judge. Just let the thoughts pass away and gently return to your breath.

• You will not be very good at letting your thoughts go, but do not worry. Whatever time you are able to spend in calm quiet—in the now, will be very helpful in reducing stress and restoring well-being.

- Even 10 or 15 minutes once or twice a day can be beneficial.

- A similar meditation can be done in a lying posture with arms and hands out to the sides.

In the early stages of Alzheimer's disease/dementia, a person could probably learn to meditate and then continue the practice with some prompts. In many ways a person with Alzheimer's disease/ dementia, spends a fair amount of time in the now, but the past memories can intrude and seem to travel in repeating patterns that can be distressing. Meditation is not about control but about gentle release of the thoughts that make our minds chatter like monkeys (many teachers call this the "monkey mind"). This practice takes some commitment but can be very beneficial. Jon Kabat- Zinn has done a great deal of work teaching meditation to chronically ill persons, especially those with cancer. His books can be helpful. Mediation is a very good way for caregivers to release worries, tension, stress and gain insight into the nature of their thoughts and feelings.

Visualization is a form of prompted meditation using a mental picture to clear the mind and focus on relaxation and healing. It is a simple technique that anyone can use to create beneficial health results. Visualization has long been used in cancer therapy to increase the cancer removing effects of the radiation and chemotherapy. All you have to do is get into a relaxed state (before you go to sleep or get out of bed in the morning are natural times) and then create a symbolic picture in your mind of illness leaving the body or healing occurring. Some people envision rivers of healing water or light surrounding their whole body or just the area that needs concentrated healing. It is not so much what symbols or pictures you use but performing the task routinely that can create the positive effects. If your person is in the early stages s/he may be able to take some calming time and visualize healing in this way. If remembering a ritual is a problem, then tapes/CDs (see the resource section in the back of this book) can be purchased, or get someone to help you create one with the images that you like. This can also be very helpful for caregivers to perform treatments on themselves for health problems and/or relieving stress related symptoms before they become chronic health issues.

Another aspect of visualization is exercises for release of worry and emotional tension. Often we carry excess anger or worry about the actions of others or guilt over our own actions. When releasing feelings over the actions of yourself or others, you focus on the action or behavior, acknowledge that it is in the past, think about any lessons you learned from what happened, make a conscious commitment to forgive, release, send a simple blessing, and move on.[22] Releasing in this way can be very helpful for your emotional health. When you cling to a vision of a particular outcome from someone else's behavior, you are wasting your energy. Excessive guilt (beyond what it takes to keep you on your personal moral path) over your own behavior also wastes energy. In truth, you do not have unlimited energy for health and healing. It is a precious commodity. In addition, you have no real control over the behavior of others only your own. Forgiving, blessing, and releasing are powerful ways to clear negative energy from your life.

CHAPTER 12
ACCEPTING AND BUILDING SUPPORT
NEVER UNDERESTIMATE THE IMPORTANCE
OF YOUR SUPPORT NETWORK

You cannot provide care for a person in your home for very long, if you are socially isolated. This isolation can increase your sense of burden, which may lead to neglect of yourself and/or the person you are caring for. Research and experience has shown that the extreme stresses of caregiving can lead to poor health and sometimes—early death in caregivers. The job of caregiving is never easy but trying to do it with little or no help or support is foolish. You have a better chance of maintaining your own health if you marshal all the resources you can to help you with the challenge of caregiving. No matter what the reason was that compelled you to become a caregiver, it is important to remain in the best possible health throughout this experience.

Each family has its own individual resources that can be coordinated into a support network. There are a number of groups you can utilize to build your own personal network. Even if you have few family members, you may be able to utilize friends and community members. The key is a bit of organization. People are often willing to help perform certain tasks that they are good at or comfortable with. The following steps will get you on your way:

- **Make a list of tasks**: You need to make a list of tasks that you could use some help with. Be flexible. It is easy to take the "I need to have it done my way" attitude. If you accept help then you need to be grateful and not critical. This does make it hard sometimes. Note that inflexibility is often a sign of high stress. Be creative about the tasks. They do not have to be limited to housework or home repairs. Having others help with recreation for your person with Alzheimer's disease/dementia can be helpful (taking your person for a walk, reading a book, or playing a game). You may have to educate your helpers on what problems your person may be currently having (especially the approximate retro-age), so the helpers are able to respond appropriately to the unexpected.

- **Make a list of available people**: Once you have adjusted your worries and feelings and have created a list of tasks, then make a list of people who may be able to help. Include individuals who may have volunteered in the past and you didn't know what to tell them. Consider family members. This chapter discusses a number of resource groups where you may find people to ask. As stated previously, asking for help can be very difficult at first. Don't be put off if people decline. It might not be the right time in their lives. Move on to another likely helper.

- **Negotiate the tasks with the people**: The best method is to match the available people to the jobs they are suited for or willing to do. Contributing help by doing a few simple tasks can make someone feel very good without burning out. This is the secret of longevity in a care/support network.

- **Formulate a schedule:** You will need to create a schedule that works for you and accommodates the times available from your volunteers. A schedule is valuable in reducing potential chaos from having a number of people enter your life. Again use flexibility in working in changes, but if it doesn't work for you say so. This plan is supposed to be a help not a hindrance.

- **Look for master helpers**: If you are lucky, you may find some people who wish to help you organize all of this more formally. If so, there are some resources that may help. Authors

Capossela and Warnock have written a book called *Share the care* giving a whole method for organizing care groups or networks for persons who are seriously ill. Although this book was not written with Alzheimer's disease/dementia in mind, it does have a lot of valuable and universal suggestions.

• **If all this seems overwhelming, start simple:** Just find one or two people to help. Use the same steps but start very small. Some help is better than none, and you may just enjoy the company. Again, social isolation is always a drain on your personal resources.

The following is a discussion of different formal and informal resource groups to help you in formulating your support network:

A. THE FAMILY/FROM SIBS TO CHILDREN TO GRANDCHILDREN

Keeping one's family intact and loving is a great achievement in life, and the great return is that they may be there for you when you need them. If this is true for you, your family will be your greatest resource. The burden if possible should not fall to one person. Everyone can help in his or her own way. This is key. Individual skills, education, and talents will vary. Tasks can be matched to skills and talents as well as desire. By spreading the helping around everyone can feel involved. As stated before, keep in mind the age and developmental level when involving children and adolescents. Children can do playful or mock work activities with your person, keeping him/her occupied while you perform other household duties. Reading is a wonderful exchange. If your person can read then this is a simple way to occupy time. If not, then have the child read to your person. You can often use the ruse of having your person help the child read the book. This can be a way to preserve dignity. In this way children can make very important contributions, especially when your person has a tendency to follow you around the house. The art of distraction and involvement are key in dealing with people with Alzheimer's disease/dementia. It helps to have people assisting you with all kinds of tasks.

Respite is another very important contribution that a large family network can provide. Your person may routinely stay in other family households so that you can have a break. Many things are possible with creative thinking. Keep in mind that your person in middle and late stages may have difficulty handling new places and may become confused. If you are concerned, do a trial overnight and see how things work out. If your person does "sundown" and wander the house or become agitated, this idea may not work for your situation. Remember each situation is **different**—not good or bad. If one idea doesn't work, try another. If overnights don't work, day trips probably will. Field trips to the zoo, arboretum, and nature sanctuary can be soothing trips that extended family can offer. These trips work very well in consort with children and grandchildren.

B. SUPPORT GROUPS

If extended family is not available or you need more help, formal systems can be accessed. The Alzheimer's Associations hold support groups in most cities. Call your city or state branch of this organization to obtain current information. Support groups provide a format for you to discuss your problems and your feelings with other people who are experiencing similar issues. The support group leaders can steer you to community resources.

The members of the support group can become part of your network. Loneliness and isolation are two major problems when you become a caregiver (especially if it is hard to get your person out of the house). Support groups are a great resource for getting you out of the house for a break. You may have to employ a sitter or find a friend who will cover this time. You will want this time away to listen to the program and participate freely in the discussions.

C. DAY CARE

Day care can be an incredibly valuable resource. Day care for adults is very similar to children's care, except that the staff orients the activities to frail adults and those with Alzheimer's disease/ dementia. Activities can range from art to dancing and field trips. Having your person in day care from 9AM to 4-5 PM can allow you to work or perform household duties uninterrupted. Most persons with Alzheimer's disease/dementia enjoy day care. They may be put off at first by the new surroundings, just as young children are. Persons with Alzheimer's disease/dementia and young children learn to adjust, and then they often find new friendships, which can be a great comfort. If you are unsure about using day care, tour some day cares and observe the activities; then choose the one you find most appropriate. This is an out of pocket cost, but many programs have sliding scales to help reduce the cost for those on a fixed income.

I remember my mother-in-law holding hands with one of her new friends just like a couple of schoolgirls. It was sweet and very reassuring. The conversations can get interesting. I remember finding my mother-in-law in deep conversation with her day care friend over the fact that her friend (an 80-yr. old with dementia) was pregnant with their favorite day care worker's baby and was marriage in order? Be prepared and go with the flow.

D. NEIGHBORHOOD AND GENERAL COMMUNITY

Your support network helpers will truly be wherever you find them. If your friends and neighbors are willing to help, go for it. Neighbors can make great sitters because they may be familiar to your person. Don't burden anyone, but take him or her up on his/her offer. Neighbors can also be great for helping with home repairs or yard work. Friends of friends may also make good helpers. Ask around and be creative. Listen to what each person is offering and make sure s/he can realistically perform the tasks, so that each situation is a win for both of you. If things don't turn out well, learn from what happened and make adjustments. Thing will work out well if you are flexible.

E. RESPITE CARE

A number of communities around the country have formal respite services. Some have income criteria and others you can simply pay for. There are even short stay programs in a nursing home that can be arranged for a fee. If a city or state runs the respite program, they may have case managers who can help you to find other resources. These programs may also offer counseling. Call your local Senior Services division, or center, or Alzheimer's Association to see if these programs are available in your area.

F. MEDICAL SUPPORT

It is invaluable to have a good medical team available to you and your person. If you have a physician that you like and trust, it is best to use him/her as the core of your medical network. This MD can refer you to a number of specialists who can provide neurological assessments and testing. If the Alzheimer's disease/dementia is in the early stages, medicinal treatments may be warranted and you will need a specialized MD to administer the drug regimen. If no drug therapy is needed, then your family doctor may be the best choice for follow-up care. These decisions can be especially important when it comes to making medical treatment and end-of-life choices.

G. HOME HEALTHCARE

Home healthcare can be a valuable resource in your support network, but this service has some limitations. There are several types of services offered under the umbrella of Home healthcare:

• **Intermittent care** Usually paid for by Medicare and insurance. The visits are short (1-2 hrs). Daily visits are not provided unless there is a specific need (usually wound or ulcer care) and a date is set when the daily care will end. These visits are for patient assessment, treatment and caregiver instruction usually by a skilled service (i.e. RN or a Physical, Occupational, or Speech therapist. This service is best utilized when your person is suffering with an acute illness or exacerbation of a chronic condition and is homebound.

• **Private Duty** Usually paid for out of pocket or by special state programs that may have income qualifications. Visits are longer—there is usually a 2- 4hr. minimum. All levels of services are offered for a fee.

• **Respite Care** This is basically a private duty program that is specifically set up to provide sitting services to allow caregivers some free time. These services are out of pocket unless there is an organization in your community that will pay the costs.

• **Hospice** This service is available for persons whose physicians will confirm that they have six months left to live. This service allows your person the freedom to die in his/her own home if s/he wishes. There are also in-patient hospice units for those who would like similar services provided in a hospital setting. Within Home healthcare there are a number of types of personnel that can be accessed based on the needs of the person with Alzheimer's disease/dementia:

• **Registered Nurses** provide skilled care services that require a nurse's level of knowledge. They can perform, for example, wound care, administer IV medication, and monitor vital signs. Licensed Practical Nurses are utilized in some states to provide similar services.

• **Home Health Aides** can assist with personal care especially bathing and provide some limited medical support. They are not allowed to provide skilled medical care and are not used for housekeeping duties.

• **Personal Care Attendants** provide personal care; prepare meals and feed your person with; and perform some housekeeping duties. They cannot provide any medical support or medical care.

• **Homemakers** provide cleaning and home support services (cooking, laundry, meal preparation, and grocery shopping); but no hands-on care. Homemaker services are usually an out-of-pocket cost unless there is a state or local program offering them. Some organizations may have a sliding scale to make services more affordable.

• **Companions** are adult sitters who usually provide respite care and companionship. They can keep your person occupied with games, stories, and conversation. They usually do not provide homemaking services at the same time (this service costs less for this reason). They can remind your person to take his/her medicines, hold an arm or hand to provide support, and prepare snack/ light meals/liquids. If there are some medical tasks that need to be performed during the respite time, a higher level of service will be needed.

• **Social Workers** are usually assigned to a case if you are having some difficulty accessing or providing needed services. They can be very helpful deciphering for you the complex world of community services. Sometimes they can provide short-term counseling. They also can be hired in the community as private counselors.

• **Physical Therapists** use exercise, massage and other techniques to provide strengthening and increased range of motion, allowing your person to return to the highest level of physical/ muscular function s/he can, after an acute illness or exacerbation of a chronic illness. Your physician will decide if your person needs this type of rehabilitation service.

• **Occupational Therapists** use special exercises and treatments to strengthen fine motor skills often involving tasks of daily living (such as eating, bathing, and dressing). They can assist in relearning skills or learning new adapting skills to increase function and self-care abilities.

• **Speech Therapists** work to correct speech and swallowing disorders especially after loss from illness, stroke, surgery, or trauma.

A person's need for skilled services (Nursing, Physical, Occupational therapy, Speech therapy, and Social work) is evaluated by the physician in conjunction with the Home healthcare staff. All the skilled services are usually short term, provided until some reasonable level of function is regained and will be discontinued when your person is no longer homebound. If rehabilitation needs persist, then outpatient therapy is recommended.

Less skilled services are freely available for a fee. It is best to ask around and get personal and professional recommendations when choosing an organization. If you cannot afford to pay for these services but feel you need them, call your state Agency on Aging or Alzheimer's Association and find out if any programs are offered on the local level.

In order to use Home healthcare and have Medicare pay for it, your person needs to be home-bound. This essentially means your person can leave home for medical appointments and sometimes day care, but not to social occasions, etc. In truth persons with Alzheimer's disease/dementia are often too physically well to use Home healthcare, even though they are mentally impaired. If there has been a recent hospitalization, Home healthcare is a very reasonable support. Your physician will usually order it from the hospital as part of discharge planning. Again, skilled Home healthcare is usually a short-term strategy to aid in post-hospital recovery.

They can also be helpful if your person is bed bound. At this point the person you are caring for may need full 24-hour care. You may or may not want to do this in your home. Many people place their loved one in a nursing facility at this point. Twenty-four hour care can be exhausting even with help. It is a decision each family must make.

Private-duty home health services can always be obtained for non-homebound persons by privately paying for the services. The services of nurses, home health aides, personal care attendants, homemakers, and companions can be purchased from many organizations. These services are expensive. If you can afford it, a personal care attendant to give a bath every few days may be a great help (if your person will cooperate). There are often a minimum numbers of hours you must purchase. Check the pricing carefully with each organization that you are considering. If the health issues are not critical, it may be your budget that dictates whether or not you utilize Home healthcare. For tips on how to work with medical organizations and medical personnel read **CHAPTER 7.**

CHAPTER 13
LONGTERM CARE PLACEMENT AND
END-OF- LIFE DECISIONS

Caregiving is a gift and devotion; but it can be an all-consuming task. So when do you draw the line? When does it all become too much? These are such difficult questions and the answers are individual to each family. Some key questions that play a factor in this decision making are:

• **What is the stage of AD of the person you are caring for?** Obviously as the person is less and less able to care for himself/herself, there is a greater burden for the caregiver. When the person with Alzheimer's disease/dementia becomes fully dependent, the care will be very consuming and too much for many people.

• **What is your own health like?** Are you well and in good physical shape or are you suffering from acute or chronic health problems? (There is a trend toward severe ill health in caregivers when the caregivers become highly stressed and sacrifice their own health needs to provide this care.)

• **Do you have any physical disabilities?** Is the level of caregiving you are providing getting to be too much for you to physically handle? Are you in danger of injuring yourself? Do your problems make it hard for you to keep your person with Alzheimer's disease/dementia safe?

• **What is the state of your emotional health?** Are you stressed and depressed? Are you taking medication to manage any mental health conditions? Depression and stress aggravate any health condition and can promote chronic health problems.

• **What amount of support do you have?** If you are fortunate to have a good family and community support network, then you will be able to stay in a home caregiving role longer. As the care gets more time consuming you can share more tasks with your network.

• **What is the condition of your support network?** Are your helpers starting to become overloaded and emotionally stressed? If they are family members, how is their physical and emotional health? If many of your helpers are also experiencing excessive stress, then this may add to the urgency of your decision.

• **What are the end-of- life decisions your person has made?** Did s/he give you any direction on when to consider placement? Did your person desire to die at home if possible? Are there any documents produced during the person's lifetime that could provide direction? These are key considerations and should be brought up during moments of lucidity especially during earlier stages of Alzheimer's disease/dementia. These decisions are also part of a general spiritual discussion you should have with your person about their beliefs and feelings surrounding the end of his/her life.

• **What is the state of your finances?** Can you afford to keep the home care that may be needed to manage your person in your home? Is placement in long-term care a more affordable and safer option?

• **How does the rest of the family feel about the decision?** Are they slanted in one direction or another? Can you get consensus or will you have to make the decision by yourself? Are there other friends and professionals you can turn to in sorting out the decision?

A. THE DECISION AND ITS EMOTIONAL IMPACT

As you can see by all the questions outlined above, this decision can become very complex. From the outside it just seems like caregivers get overwhelmed and then their person with Alzheimer's disease/dementia is placed in a facility and that is the end of it. This decision is neither simple nor easy. Even with direction and support from family and friends there will be a grieving process. You have been grieving little by little all through your time of caregiving, especially at each juncture where your person has lost more self-care abilities. Placement is a large milestone in the grief process. You will grieve before, during and after you have made the decision and when your person has been moved. It will at times be mixed with relief that your time of intense giving of your physical self is over. Your time of emotional giving will of course continue.

This is a very important time to seek emotional support and professional help if needed. You will need to accept your feelings no matter how wide-ranging. You may feel happy and sad at the same time. You may experience great guilt. Perhaps you promised your person placement would never happen (It is never a good idea to do this because you cannot predict the future and the state of your own health). Guilt is sometimes useful to tweak us when we have strayed from our values, but it should not be used to beat yourself up. If you feel guilt, examine the source, perhaps you need to do some emotional work and adjust your values. They may no longer suit your current life situation. Even as adults, we all grow as we live; and our life experiences present challenges that spur on further growth. This is a wonderful dimensional process and one that will bring many rewards. If we try to escape into avoidance or depression, the process will be stopped short and bring more anguish. We all have to move through our pain into growth. Losing loved ones is a way-crossing in life and may happen many times. If you live in your grief and work through it, you will emerge transcendent. Be kind to yourself and forgive.

B. WHERE TO PLACE AND THE QUALITY OF CARE

Once the decision to place your person in long-term care has been made, then you have to make the difficult choice as to where you would like him/her to live. There are a number of considerations:

• **What level of care does your person need?** If your person requires full care with little or no ability to toilet himself/herself or is bed bound, you will need nursing home care. If s/he is still able to provide some self-care, assisted living is more appropriate. In the least restricted settings there are more activities and entertainment for the more active person with Alzheimer's disease/dementia. Many institutions now have several levels of care. This is very helpful because your person can move from one level to the next, as his/her needs increase.

• **Are there attractive places near where you live?** This is helpful because it cuts down commutes and makes visiting easier. This is especially important if you no longer drive and have to depend on others for transportation. It may be good to survey all nearby places first and then move out in ever widening circles to look at other places.

• **What do you consider attractive in a supported living situation?** Think about if for a bit. Make a list or jot down ideas. Take a variety of tours. This will give you some concrete ideas of what to expect. Observe a group activity to see if they seem engaging and pleasurable for the residents. Observe the people working at each facility. Do they seem friendly and helpful? Is there an unpleasant smell in the facility? Most nursing facilities work hard to minimize smell, so this is less of a problem than in the past. If you are seriously considering a facility then speak to the director about your concerns.

• What are their visiting rules? Most modern institutions try to accommodate family especially spouses, but make sure. Would you be allowed to stay over if there are problems/health crises?

• What is your person's view about end-of-life decisions? Is it compatible with yours? Will your person with Alzheimer's disease/dementia want you to withhold treatment if you and your physician agree it is the right thing to do? Do they allow access to hospice services if you would like them? It is good to be as clear as possible as to your desires so that conflicts do not arise.

This is just a brief overview of some of the components of making a placement decision. The Alzheimer's Association as well as state and local Agencies on Aging often have pamphlets that go over all of the critical elements of nursing home and assisted living quality to help you choose the best placement for your person.

C. END-OF-LIFE DECISIONS

Alzheimer's disease/dementia can present a number of end-of-life decision challenges. Much is written about the various legal powers you need to obtain such as "power of attorney" for financial and/or health related matters. Once you have this decisional power then you may actually be called upon to use it. It is best to be prepared for this time. You cannot be entirely ready because you never know what will happen to your person, but you can discuss the issues with him/her if you have the opportunity and with family and friends. Advanced directive packets sometimes come with forms for "values clarification." These forms are not required in any legal sense, but they are very helpful for yourself and the person you are caring for. They help you clarify what the person with Alzheimer's disease/dementia feels, thinks, and values about living and dying. You can identify differences and/or similarities in the way you both think and feel about these matters. If there is a moment of lucidity, you can even get firm direction on issues such as the prolongation of life, cremation, burial, and funeral arrangements. If it is too late to get firm direction, see what can be clarified, and add it to what you know about your person's former desires; and discuss the matters with close family members and friends.

If you have been given decisional responsibility, then you may have to make some difficult decisions. Hopefully you will have supportive family and friends to give you love and guidance. Issues around the prolongation of life are some of the hardest and most conflicting. Most people are now aware of ventilators and can make the call that they do not want be put on one if they are determined "brain dead." In Alzheimer's disease/dementia, this is hardly ever the call you will have to make. You are more likely to have to decide about providing or withholding other treatments such as artificial nutrition, surgeries, and special medications.

In End Stage Alzheimer's disease/dementia, a person essentially goes back to an infant/fetus. If your person has other diseases then s/he may have a health crisis and die of complications of those diseases. If not, then Alzheimer's disease/dementia will follow its course and at some point the person will not be able swallow and will have to have artificial nutrition to prolong his/her life. At this point all the complex issues of caring for a non-responsive bed bound individual come into the picture, including trying to prevent skin ulcers. This type of care with artificial nutrition is very consuming and you may not be able to handle it in your home. Your physician and support network will be very important at this point in helping you make decisions that are compatible with your values and those of your person. Hospice can also be of great help as they are very skilled at supporting people through the physical, emotional, and spiritual aspects of making end-of-life decisions. No book can replace good spiritual counseling at this time, even if you are not religious. Most hospice services are non-denominational, but they can put you in touch with chaplains or ministers of the faith/belief of your choice. Do not overlook your close friends for counsel.

Everyone has to face death, and giving support can help clarify values and spiritual beliefs of all that come to help. These decisions are part of our life and humanity. We are strengthened by these challenges not damaged or weakened by them. Losing loved ones is all part of life, rarely escapable unless one dies at a young age. Each time we support someone in their dying process, we are immersed in the lessons of spiritual intensity and love. Hospice workers and volunteers stay involved because of this intensity and connection. Witnessing death and birth is equally powerful, as they are both natural parts of the spectrum of life. In our society, which has "medicalized" both of these human moments, it is a rare privilege and a blessing to be able witness either one. If you find yourself immersed in end-of-life decision making, look to your own soulful compassion, these perspectives, and the help of your medical, social and spiritual support network to make this time one of the most humane, human and beloved of moments in this, your life.

ENDNOTES

CHAPTER 8

[1] Barry Reisberg et al., "Evidence and mechanisms of retrogenesis in Alzheimer's and other dementias: management and treatment import," *American Journal of Alzheimer's Disease and Other Dementias* 17 (2002): 202-12.

[2] Howard Gardner, *Multiple INTELLIGENCE: the Theory in Practice, a reader* (New York: BasicBooks, 1993) 15.

[3] Ibid, 17-25

[4] Kenneth Mitchell, and Herbert Anderson, *All Our Losses, All Our Griefs: Resources for Pastoral Care* (Louisville: Westminster John Knox Press, 1983) 61-82.

CHAPTER 10.

[1] Carolyn Myss, & C. Norman Shealy, *The Creation of Health: The Emotional, Psychological, and Spiritual Responses That Promote Health and Healing.* (New York: Three Rivers Press, 1993), 294-9.

[2] Ibid, 298-9.

[3] Elmer Green, *The Ozawkie Book of the Dead: Alzheimer's Isn't What You Think It Is!* (Vols. 1-3) (Los Angeles: The Philosophical Research Society, 2001).

[4] Sogyal Rinpoche, *The Tibetan Book of Living and Dying.* (San Francisco: Harper, 1992), 102.

[5] J. Kaye, & K. M. "Spirituality Among Caregivers," *Image Journal of Nursing Scholarship,* 26(3) (1994): 218-21.

[6] Green, *The Ozawkie Book of the Dead,* 1:2.

[7] Ibid., 1: 4.

[8] Evelyn Underhill, *The Spiritual Life* (Atlanta: Ariel Press, 2000).

[9] Carolyn Myss, *Sacred Contracts: Awakening Your Divine Potential* (New York: Harmony Books, 2001), 407.

[10] Richard J. Foster, *Celebration of Discipline: The Path to Spiritual Growth* (San Francisco: Harper Collins, 1998), 130.

[11] Ibid., 132.

[12] Lama Surya Das, *Awakening The Buddha Within* (New York: Broadway Books, 1997), 157.

[13] Rinpoche, *The Tibetan Book of Living and Dying,* 102.

[14] Ibid., 187.

[15] Pema Chodron, "The In–Between State," *Tricycle* 42 (2001): 56.

[16] Rudolph Ballentine, *Radical Healing* (New York: Three Rivers Press, 1999), 498.

[17] Foster, *Celebration of Discipline,* 33.

[18] Virginia Morris, *How to Care for Aging Parents* (New York: Workman Publishing, 1996), 42.

[19] Larry Dossey, *Healing Words:The Power of Prayer and the Practice of Medicine* (San Francisco: HarperCollins, 1993), 97.

[20] Ibid., 98.

[21] Timothy Freke, *Spiritual Traditions: Essential Teachings to Transform Your Life* (New York: Sterling Publishing Company, 2001), 66.

CHAPTER 11.

[1] *The Complete Family Guide to Alternative Medicine: An Illustrated Encyclopedia of Natural Healing,* ed. C. Norman Shealy and Richard Thomas (Rockport, Massachusetts: Element Books Limited, 1996), 38.

[2] Ibid., 42.

[3] Elizabeth Kaledian, *New Hope for Alzheimer's* (New York: CBS evening News, 2002).

[4] Porter Shimer, *New Hope: For People with Alzheimer's and Their Caregivers* (Roseville, California: Prima Publishing, 2002), 97.

[5] Ibid., 98.

[6] Shealy, *Family Guide to Alternative Medicine*, 52.

[7] *The Complete Illustrated Guide of Alternative Healing Therapies,* ed. C. Norman Shealy (New York: Barnes&Noble Books, 1996), 356.; Burton Goldberg, *Alternative Medicine: The Definitive Guide*, ed. Larry Trivieri and John W. Anderson (Berkeley: Celestial Arts, 2002), 530.

[8] Shealy, *Family Guide to Alternative Medicine*, 52.

[9] Goldberg, *The Definitive Guide*, 526.

[10] Robert C. Atkins, *Dr.Atkins' Vita-Nutrient Solution* (New York: Fireside Books, 1999), 336.; Goldberg, *The Definitive Guide*, 530.

[11] Shealy, *Family Guide to Alternative Medicine*, 90.

[12] Goldberg, *The Definitive Guide*, 530.

[13] Shimer, *New Hope*, 94.

[14] Don Campbell, *Music Physician for Times To Come* (Wheaton, Illinois: Quest Books, 1991), 1-8.

[15] Shimer, *New Hope*, 94.

[16] Ibid., 95.

[17] C. Norman Shealy, *The Complete Illustrated Guide of Alternative Healing Therapies*, 210.

[18] Goldberg, *The Definitive Guide*, 529.

[19] C. Norman Shealy, *Sacred Healing: The Curing Power of Energy and Spirituality* (Boston: Element, 1999), 175.

[20] Goldberg, *The Definitive Guide*, 326-9.

[21] Karen Grace Kassy, *Health Intuition: A Simple Guide to Greater Well-Being* (Center City, Minnesota: Hazelden, 2000), 52.

SELECTED READINGS

ALZHEIMER'S DISEASE AND GENERAL CARE

Baley, J. 1999. *Elegy for Iris.* New York: St. Martin's Press.

Capossela, C. & Warnock, S. 1995. *Share the Care: How to Organize a Group to Care for Someone Who is Seriously Ill.* New York: Fireside Books.

Carroll, D. L. 1989. *When Your Loved One Has Alzheimer's: a Caregiver's Guide.* New York: Harper & Row.

Cohen, D. & Eisdorfer, C. 2001. *The Loss of Self: A Family Resource for the Care of Alzheimer's Disease and Related Disorders.* New York: W. W. Norton & Company.

Coste, J. K. 2003. *Learning to Speak Alzheimer's: A Ground Breaking Approach for Everyone Dealing with the Disease.* New York: Houghton Mifflin Company.

Davidson, F. G. (1993). *The Alzheimer's Sourcebook for Caregivers: A Practical Guide for Getting Through the Day.* Los Angeles: Lowell House.

FitzRay, B. J. 2001. *Alzheimer's Activities: Hundreds of Activities for Men and Women With Alzheimer's Disease and Related Disorders.* Windsor, California: Rayve Productions.

Gruetzner, H. 2001. *Alzheimer's: A Caregiver's Guide and Sourcebook.* New York:John Wiley & Sons.

Hodgson, H. (1995). *Alzheimer's Finding the Words: A Communication Guide for Those Who Care.* New York: John Wiley and Sons.

Hodgson, H. (1995). *The Alzheimer's Caregiver: Dealing with the Realities of Dementia.* Minneapolis, MN: Chronimed Publishing.

Mace, N. L. & Rabins, P. V. (1981). *The 36-hour Day: A Family Guide to Daring for Persons with Alzheimer's Disease, Related Sementing Illnesses, and Memory Loss in Later Life.* New York: Warner Books.

Morris, V. 1996. *How to Care for Aging Parents.* New York: Workman Publishing.

Reisberg, B. (2002). *Evidence and Mechanisms of Retrogenesis in Alzheimer's and Other Dementias: Management and Treatment Import.* American Journal of Alzheimer's Disease and other Dementias, 17 (4), 202-12.

Shanks, L. K. (1996). *Your Name is Huges Hanibal Shanks: A Caregiver's Guide to Alzheimer's.* New York: Penguin Books.

Shenk, D. (2001). *The Forgetting: Alzheimer's Portrait of an Epidemic.* New York: Doubleday.

Shimer, P. (2002). *New Hope for People with Alzheimer's and Their Caregivers.* Roseville, CA: Prima Publishing.

Snowdon, D. 2001. *Aging with Grace: What the Nun Study Teaches Us About Leading Longer, Healthier and More Meaningful Lives.* New York: Bantam Books.

Thomas, W. H. 1996. *Life Worth Living: How Someone You Love Can Still Enjoy Life in Nursing Home-the Eden Alternative in Action.* Acton, Massachusetts: VanderWyk & Burnham.

EDUCATION

Gardner, H. (1993). *Multiple INTELLIGENCE: The Theory in Practice.* New York: BasicBooks

THE BRAIN

Austin, J. H. 1998. *Zen and the Brain.* Cambridge: MIT Press.

Newberg, A., D'Aquili, E., & Rause, V. (2001). *Why God Won't Go Away.* New York: Ballentine Books.

Ramachandran, V. S. & Blakeslee, S. 1998. *Phantoms in the Brain: Probing the Mysteries of the Human Mind.* New York: William Morrow and Company.

Sacks, O. 1985. *The Man Who Mistook His Wife for a Hat and other Clinical Tales.* New York: HarperCollins.

HEALING/ENERGY MEDICINE

Achterberg, J., Dossey, B., & Kolkmeier, L. 1994. *Rituals of Healing: Using Imagery for Health and Wellness.* New York: Bantam Books.

Advice on Dying: By His Holiness the Dali lama. 2002. Ed. Jeffery Hopkins. New York: Atria Books.

Atkins, Robert C. *Dr.Atkins' Vita-Nutrient Solution. 1999* New York: Fireside Books.

Ballentine, R. 1998. *Radical Healing.* New York: Three Rivers Press.

Campbell, D. 1991. *Music Physician for Times To Come.* Wheaton, Illinois: Quest Books.

Dossey, L. 1993. *Healing Words.* San Francisco: Harper & Row.

Goldberg, B. *Alternative Medicine: The Definitive Guide.* 2002. Ed. Larry Trivieri and John W. Anderson. Berkeley: Celestial Arts.

Kabat-Zinn, J. 2005. *Full Catastrophe Living: Using the Wisdom of Your Body and Mind to Face Stress, Pain, and Illness.* New York: Bantam Dell.

Kassey, K. G. 2000. *Health Intuition: A Simple Guide to Greater Well-Being.* Center City, Minnesota: Hazelden.

Myss, C. M. 1997. *Why People don't Heal and How They Can.* New York: Three Rivers Press.

Myss, C. M., & Shealy, C. N. (1993). *The Creation of Health: The Emotional, Psychological, and Spiritual Responses That Promote Health and Healing.* New York: Three Rivers Press.

Naparstek, B. 1994. *Staying Well With Guided Imagery: How to Harness the Power of Your Imagination for Health and Healing.* New York: Warner Books.

Page, C. & Hagenbach, K. 1999. *Mind, Body, Spirit Workbook: A Handbook of Health.* Saffron Walden, UK: The C.W. Daniel Company Limited.

Pert. C. B. 1997. *Molecules of Emotion: The Science Behind Mind-Body Medicine.* New York: Simon and Schuster.

Schulz, M. L. 1998. *Awakening Intuition: Using Your Mind-Body Network for Insight and Healing.* New York: Three Rivers Press.

Shealy, C. N. 1999. *90 Days to Stress-Free Living: A Day by Day Health Plan Including Exercises, Diet and Relaxation Techniques.* Boston: Element.

Shealy, C. N. 1995. *Miracles Do Happen: A Physician's Experience With Alternative Medicine.* Rockport, Massachusetts: Element.

The Complete Family Guide to Alternative Medicine: An Illustrated Encyclopedia of Natural Healing. 1996. Ed. C. Norman Shealy and Richard Thomas. Rockport, Massachusetts: Element Books Limited.

The Complete Illustrated Guide of Alternative Healing Therapies, 1996. Ed. C. Norman Shealy. New York: Barnes & Noble Books.

SPIRITUALITY AND INSPIRATION

Borang, K. K. 2001. *Way of Reiki*. London: Thorsons.

Chodron, P. 1997. *When Things Fall Apart: Heart Advice for Difficult Times*. Boston: Shambala.

Forrest, D. A. 2000. *Symphony of Spirits: Encounters with the Spiritual Dimensions of Alzheimer's*. New York: St. Martin's Press.

Foster, R. J. 1998. *Celebration of Discipline: The Path to Spiritual Growth*. San Francisco: HarperCollins.

Frankl, V. E. 1984. *Man's Search for Meaning: An Introduction to Logotherapy*. New York: Simon & Schuster.

Freke, T. 2001. *Spiritual Traditions: Essential Teachings to Transform Your Life*. New York: Sterling Publishing Co.

Green, E. 2001. *The Ozawkie Book of the Dead: Alzheimer's Isn't What You Think It Is! Vols. 1-3*. Los Angeles: The Philosophical Research Society

Kubler-Ross, E. 1986. *Death: The Final Stage of Growth*. New York: Touchstone.

Levin, J. 2001. *God, Faith, and Health: Exploring the Spirituality-Healing Connection*. New York: John Wiley & Sons.

Mother Teresa's Reaching Out in Love. 2002. Ed. Edward Le Joly and Jaya Chaliha. New York: Barnes & Noble Books.

Muller, Wayne. 1999. *Sabbath: Finding Rest, Renewal, and Delight in Our Busy Lives*. New York: Bantam Books.

Myss, C. M. 1996. *Anatomy of Spirit: The Seven Stages of Power and Healing*. New York: Three Rivers Press.

Myss, C. M. 2004. *Invisible Acts of Power: Personal Choices That Create Miracles*. New York: Free Press.

Myss, C. M. 2001. *Sacred Contracts: Awakening Your Divine Potential*. New York: Harmony Books.

Naparstek, B. 1997. *Your Sixth Sense: Unlocking the power of Your Intuition.* San Francisco: Harper.

Newberg, A., D'Aquili, E., & Rause, V. 2001. *Why God Won't Go Away*. New York: Ballentine Books.

Rinpoche, S. 1992. *The Tibetan Book of Living and Dying*. San Francisco: Harper.

Schachter-Shalomi, Z. 1997. *From Age-ing to Sage-ing: A Profound New Vision of Growing Older.* New York: Warner Books.

Shealy, C. N. 1999. *Sacred Healing: The Curative Power of Energy and Spirituality*. Boston: Element.

Lama Surya Das. 1997. *Awakening The Buddha Within*. New York: Broadway Books.

Teasdale, W. 1999. *The Mystic Heart: Discovering a Universal Spirituality in the World's Religions*. Novato, California: New World Library.

Underhill, E. 2000. *The Spiritual Life*. Atlanta: Ariel Press.

RESOURCES

For Health preparations and tapes by C. Norman Shealy MD
Self-Health Systems
5607 S. 222nd Road
Fair Grove, MO 65648
(417) 267-3102

For specialty items
The Alzheimer's Store
www.alzstore.com

For Visualization tapes and CDs
Go to the website for Belleruth Naparstek
Health Journeys: Resources for Mind, Body and Spirit
www.healthjourneys.com

Don't forget to visit the Alzheimer's Workbook Website and register.
www.alzheimersworkbook.com